Routledge Revivals

Indian Hirudinea
(RHYNCHOBDELLÆ)

Indian Hirudinea
(RHYNCHOBDELLÆ)

by

W. A. Harding

Routledge
Taylor & Francis Group

First published in 1927 by Taylor & Francis

This edition first published in 2019 by Routledge
2 Park Square, Milton Park, Abingdon, Oxon, OX14 4RN
and by Routledge
52 Vanderbilt Avenue, New York, NY 10017

Routledge is an imprint of the Taylor & Francis Group, an informa business

© 1927 by Taylor & Francis

Publisher's Note
The publisher has gone to great lengths to ensure the quality of this reprint but points out that some imperfections in the original copies may be apparent.

Disclaimer
The publisher has made every effort to trace copyright holders and welcomes correspondence from those they have been unable to contact.
A Library of Congress record exists under ISBN:

ISBN 13: 978-1-138-31846-5 (hbk)
ISBN 13: 978-0-429-45456-1 (ebk)

Plate II.

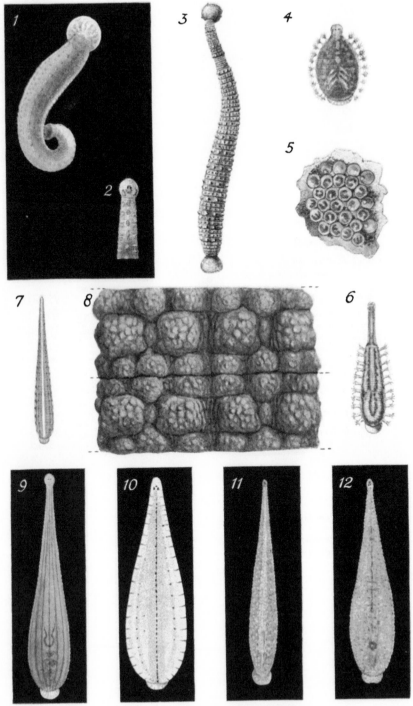

INDIAN RHYNCHOBDELLÆ.

EXPLANATION OF THE PLATE.

INDIAN RHYNCHOBDELLÆ.

Fig. 1. *Hemiclepsis marginata*, subsp. *marginata* (O. F. Müller), 1774. An individual gorged with blood, hanging by its posterior sucker. × 14.

Fig. 2. The same. Dorsal view of anterior part of the body. × 14.

Fig. 3. *Pontobdella loricata*, Harding, 1924. From a specimen preserved in alcohol, life size.

Fig. 4. *Ozobranchus shipleyi*, Harding, 1909. A dark-coloured example in contraction. × 2.

Fig. 5. *Ozobranchus shipleyi*. A patch of eggs embedded in chitinous cement, stripped from the plastron of the tortoise, *Kachuga intermedia*, viewed by transmitted light as a transparency. × 7.

Fig. 6. *Ozobranchus shipleyi*. Dorsal view of a young individual, fully extended. × 3.

Fig. 7. *Placobdella fulva*, Harding, 1924. Dorsal aspect. × 3.

Fig. 8. *Pontobdella* (subgen. *Pontobdellina*) *macrothela*, Schmarda, 1861. Dorsal aspect of somites XIX and XX, showing the disposition of the tubercles. × 5.

Fig. 9. *Helobdella nociva*, Harding, 1924. Dorsal aspect. × 8.

Fig. 10. *Glossiphonia weberi*, R. Blanchard, 1897. Dorsal aspect. × 8.

Fig. 11. *Paraclepsis prædatrix*, Harding, 1924. Dorsal aspect. × 5.

Fig. 12. *Paraclepsis prædatrix*. Dorsal aspect of another individual of a different colour. × 5.

N.B.—Figs. 1, 2, 4–6, and 9–12 are by A. C. Chowdhary, of the Indian Museum, and figs. 3 and 8 by the late W. West, of Cambridge.

SYSTEMATIC INDEX.

INTRODUCTION

TO THE

RHYNCHOBDELLÆ.

BY

W. A. HARDING, M.A., F.L.S.

THE Suborder Rhynchobdellæ comprises leeches which suck the blood and juices of their prey by means of a protrusible proboscis. It contains no terrestrial or carnivorous forms. Its members are strictly parasitic, and consist, without exception, of freshwater and marine species, the latter being the only leeches known to inhabit the sea.

The proboscis, which constitutes so important a characteristic of this group, is an adaptation of the pharynx, which has become highly muscular, free and protractile, and so capable of being thrust through the small oral opening in the leech's anterior sucker into the tissues of its host.

Few Rhynchobdellæ are large, and none reach the formidable proportions attained by certain predaceous Arynchobdellæ. In the Indian region, where they are widely distributed, they are all, with the exception of a few marine species, of small size, ranging from about 6 mm. to 20 mm. in length, and on this account, combined with the fact that they are innocuous to mankind, they do not generally attract the attention of the ordinary observer.

The Suborder is divided into the three Families—Acanthobdellidæ (not represented in India), Ichthyobdellidæ and Glossiphonidæ. It must be noted, however, that opinion is not unanimous in regarding *Acanthobdella peledina*, the only representative of the first of these Families, as a leech. Notwithstanding its affinities with the Hirudinea, some authorities consider that its true position is with the Oligochæts. In the Rhynchobdellæ there is always a permanent, cupuliform or discoid posterior sucker, directed more or less ventrally and supported, as on a kind of pedicel, by the tapering portion of the body lying immediately behind it. An anterior sucker of very similar form is nearly always found in the Ichthyobdellidæ, but in a few members of this Family and in the Glossiphonidæ this is of a different type.

Here the head-region of the body preserves its annulation dorsally to its anterior end, and the sucker takes the form of a more or less spoon-shaped depression hollowed out of its ventral surface.

The suckers, which are highly muscular, form powerful organs of attachment, and the extent of their development bears some relation to the activity of the hosts to which they adhere. Suckers showing but a moderate development are seen in the Glossiphonidæ, which prey for the most part upon hosts which are sluggish or only attacked when in a state of inactivity. The Ichthyobdellidæ, however, are very largely parasitic upon fish, and so need the large suckers developed at the extremities of the body in order to remain firmly fixed to their vigorous carriers. So strong is the action of the anterior sucker of *Pontobdella* that its projecting rim frequently leaves a deeply-indented circular scar upon the body of the host.

The Rhynchobdellæ exhibit considerable diversity of form, which is most noticeable, however, in the Ichthyobdellidæ. In this Family the body, which is nearly always elongate, may be cylindrical, flattened or claviform, preserving either an unbroken outline or, on the other hand, being distinctly divided into a short anterior and a long posterior region. It may present all gradations between a nearly smooth surface and one covered by large and prominent warty tubercles, and in the case of the two genera *Ozobranchus* and *Branchellion* it bears conspicuous lateral *branchiæ* or gills. Another curious and not uncommon feature affecting, however, only temporarily the contour of the body consists in what are known as *pulsating vesicles*. These are small inflations of the skin, resembling hemispherical blisters, which rise and disappear rhythmically. They are due to a mechanism which will be referrep to later, and occur metamerically in lateral pairs on the posterior two-thirds of the body. *Pterobdella amara* (Kaburaki, 1921) affords an example of unusual configuration in an Indian species, the more or less cylindrical body being provided with paired lateral processes bearing some resemblance to fins.

Compared with the Ichthyobdellidæ, the Glossiphonidæ are a much more homogeneous assemblage of leeches, and usually show little variation from the typical flattened, ovate-acuminate form.

The Rhynchobdellæ possess a very wide range of hosts, and its members do not usually confine themselves to a particular species or even genus, but are somewhat more catholic in their tastes. The Glossiphonid species *Helobdella stagnalis*, for example, although parasitic chiefly upon aquatic Gasteropods, is not averse to various other freshwater invertebrates; it will suck out the whole contents of aquatic larvæ, and may have recourse, on occasion, to frogs, newts and injured fish. The smallest Rhynchobdellæ are able to pierce epithelial surfaces, and have been found (in Africa) in such diverse situations as the anus of an elephant, the mouth of a crocodile and the pouch of a pelican.

The Ichthyobdellidæ are parasitic chiefly upon marine and freshwater fish, of which probably few species, from sharks, skates and rays to the smallest forms, are not actual or possible hosts. Marine and freshwater Chelonia and sea-crabs are frequently attacked. Pantopodæ and a species of *Sclerocrangon* have also been recorded as victims. Certain hosts of the Glossiphonidæ have already been referred to. In addition to Gasteropods and a large number of freshwater invertebrates, including beetles (Dityscidæ) and insect larvæ, the members of this family have been found upon amphibia, aquatic birds and freshwater fish and crabs.

Although the Rhynchobdellæ have not hitherto attained notoriety as dangerous parasites and do not, as already noted, draw blood from Man, they deserve nevertheless close investigation with a view to ascertaining if they play any part in the transmission of diseases of economic importance. In this connection reference may be made to the work of Miss Robertson (1909, 1910, 1911), who has established, for example, *Ozobranchus shipleyi* as the intermediate host of a Hæmogregarine found in the blood of the lake-tortoise *Nicoria trijuga* in Ceylon, and shown that *Hemiclepsis marginata marginata* and the European leech *Pontobdella muricata* transmit Trypanosomes to certain fish.

EXTERNAL CHARACTERS.—Turning now to the consideration of external features, of which a knowledge, until comparatively recent times, was regarded as sufficient for the work of the systematist, I find my task lightened by a chapter on the segmentation of the Hirudinea, contributed to this volume by Prof. J. Percy Moore, whose name will always be remembered in connection with this subject, which he has developed still further in his introduction to the Arhynchobdellæ, in these pages. It is unnecessary, therefore, to do more here than summarise briefly the more important facts connected with the external metamerism.

The body of a leech is always composed of thirty-four segments or somites, and inasmuch as the presence of a ganglion is the fundamental test of a somite, so there are thirty-four ganglia in the central nervous system. Of these, the circum-pharyngeal ganglionic mass contains six, and the posterior ganglionic mass contains seven fused ganglia, and twenty-one free single ganglia lie in the ventral chain between them.

In the leech, again, the number of *rings* or *annuli* exceeds the number of somites, and throughout the greater part of its length these rings resolve themselves into a series of regularly recurring groups, corresponding to the successive somites of the body. Each of these similar groups containing an equal number of rings is termed in leech nomenclature a *complete somite*, and this, in the Rhynchobdellæ, may include from two to fourteen rings. Towards the ends of the body the number of rings in a group becomes progressively smaller, forming what are regarded as reduced or *incomplete somites*, and at the anterior extremity one or more somites may be represented only by a single ring.

The sensitive body of a leech is covered by minute sense-organs, but we need only concern ourselves here with the segmental sense-organs or *sensillæ*. These are confined to the sensory ring of each somite, which in the middle part of the body overlies a ganglion of the ventral chain, and appear in strict series in definite longitudinal rows or lines. According to Livanow (1903) these lines occur on the dorsal and ventral surface in pairs with respect to a median line, and counting outwards from the median line, consist typically of (1) an inner and (2) an outer paramedian pair; (3) an intermediate pair; (4) an inner and (5) an outer paramarginal (or submarginal) pair; and finally (6) of a marginal pair coinciding with the edges of the body.

It must not be supposed, however, that sensillæ are always present on these lines in every species. Generally the number of pairs is reduced, and this reduction is noticeable in the Rhynchobdellæ. The sensillæ, which often appear as small white spots on the surface of the body, are frequently associated with colour-markings and more or less prominent cutaneous papillæ, and thus the sensory rings upon which they are borne are in most cases easily recognised by the eye.

Whitman and many others regarded these conspicuous sensory rings as the first rings of the successive somites of the body. Although, however, this method of determining somite limits was convenient, it presented many difficulties which need not be discussed here, and in 1900 Prof. J. Percy Moore and Dr. W. E. Castle, each working independently and upon different material, suggested a new method of somite delimitation based upon the nervous distribution, which they very properly regarded as of fundamental importance.

Under this system, which is now generally accepted, each somite is innervated by the ventral ganglion it contains. The sensory ring lies in the middle of the three- or five-ringed somite, and takes its place there as the primary ring from which the others have been derived by a process of growth and subdivision.

The sensillæ, which are retractile and provided with sensory hairs, in addition to functioning as tactile organs, appear to some extent to be sensitive to light. On the dorsal surface of the head-region they are often specially developed and modified into eyes, which consist essentially of a nerve axis surrounded by visual cells, surmounted by an epithelial cap and embedded posteriorly in a dense black pigment-cup, which, when viewed from above, is of more or less crescent-like form. Such organs can do no more than distinguish between light and darkness.

From one to four pairs of eyes may be present in the Rhynchobdellæ. In a few cases, as in some species of *Placobdella*, eyes which at first appear to be single are found on closer examination to be compound, and occasionally the typical number may be incomplete or even exceeded; nevertheless it cannot be denied that the eyes, both in number and arrangement, are of great diagnostic value.

In addition to the number of rings, the number and constitution of the somites, complete and incomplete, the colour, markings, papillæ, eyes, vesicles, branchiæ, and the form of the body and its suckers, the openings in the body have still to be considered by the systematist. The position of the genital apertures in the ventral median line, and also that of the mouth-opening in the anterior sucker, are taxonomically of great importance; the situation of the anus in the dorsal posterior region is another useful diagnostic feature in the external topography, and the number and position of the nephridiophores upon the ventral surface, although often obscured in preserved material, should if possible be ascertained. The genital openings generally lie within the limits of the eleventh and twelfth somites, the male anterior to the female; occasionally a single slit-like pore encloses the apertures of both.

The mouth-opening, through which the proboscis is protruded, usually lies well within the cup of the anterior sucker, but sometimes it occurs at, or close to, its anterior extremity, thus leaving the interior face of the sucker imperforate. *Ozobranchus* and *Paraclepsis* afford examples of subterminal oral openings; a terminal opening is characteristic of *Placobdella*.

BODY-CAVITY OR CŒLOM.—The most notable and perplexing internal feature in the Ichthyobdellidæ and Glossiphonidæ is the cœlom. In the nearest allies of the Hirudinea, namely the Oligochæta, this consists of a fairly spacious body-cavity, divided intersegmentally by septa and containing the viscera; but among leeches this condition is seen only in the Siberian species *Acanthobdella*, which forms a connecting-link between the two groups. In the rest of the Rhynchobdellæ the body-cavity is split up into a series of longitudinal canals of varying size, connected by a complicated system of intercommunicating branches, which has been called the *lacuna system*.

Apart from the lacuna system, which is filled by the cœlomic fluid, or lymph, and entirely unconnected with it, there is present a vascular system containing colourless blood.

In typical species of the Glossiphonidæ, Oka (1894) finds that the lacuna system comprises five longitudinal trunks, namely (1) a median lacuna, (2 and 3) a pair of lateral lacunæ, and (4 and 5) a pair of intermediate lacunæ lying between them and connecting the median and lateral trunks by means of transverse lacunæ. In addition to these, a system of hypodermal lacunæ lie beneath the skin.

The median lacuna, which contains a series of imperfect septa, traverses the body from the head-region to the anus, and where the stomach and intestine occur becomes divided into a dorsal and a ventral lacuna, the two being separated by the portion of the alimentary tract referred to. The narrow dorsal lacuna encloses the dorsal blood-vessel, and the larger ventral lacuna contains the ventral blood-vessel, the nerve-cord and part of the reproductive organs.

C

The lateral lacunæ lie at the margins of the body, and also extend from the head-region to the anus. At both extremities they are connected with the median lacuna and each other by means of a circular canal passing round the margins of the anterior and posterior suckers. These lacunæ are not contractile like the lateral lacunæ of the Ichthyobdellidæ, and the cœlomic fluid in the Glossiphonidæ probably owes its circulation partly to the movements of the animal and partly to the pulsations of the dorsal blood-vessel in the median lacuna.

The intermediate lacunæ extend from the sixth somite to the anus, and, together with the transverse lacunæ, may be regarded

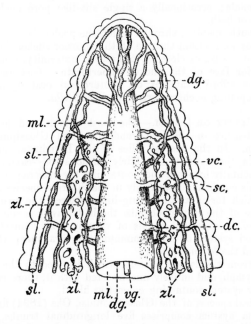

Fig. 2.—Schematic representation of the lacuna system in the anterior part of the body of *Glossiphonia complanata* (after Oka). *m.l.*, median lacuna; *zl.*, intermediate lacuna; *sl.*, lateral lacuna; *d.c.*, *v.c.*, and *s.c.*, dorsal, ventral and lateral transverse lacunæ; *dg*, dorsal blood-vessel; *vg*, ventral blood-vessel.

merely as extensions of the median lacuna. They are not simple trunks, but consist of a continuous network of canals, which occupy the spaces between the dorso-ventral muscles, nephridial cells and the connective tissue of the body.

The hypodermal lacunæ encircle the body immediately beneath the skin, and communicate with the lateral and intermediate lacunæ. From one to four of these fine canals may occur in each annulus, and by bringing the cœlomic fluid so near to the exterior surface they assist in the process of respiration.

In the Ichthyobdellidæ the lacuna system is much more variable in plan than in the Glossiphonidæ, and the lateral lacunæ, which are rarely absent, are strongly contractile, with powerful muscular

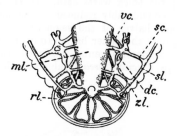

Fig. 3.—Schematic representation of the lacuna system in the posterior part of *Glossiphonia complanata* (after Oka). *m.l.*, median lacuna; *zl.*, intermediate lacuna; *sl.*, lateral lacuna; *d.c.*, *v.c.* and *sc.*, dorsal, ventral and lateral transverse lacunæ; *rl.*, circular lacuna in posterior sucker. Bloodvessels indicated by dotted shading.

walls. The system attains its highest degree of complexity in those forms which, like *Piscicola geometra*, are provided with *pulsating vesicles*.

In this species, which has been investigated by Johansson (1896) and Selensky (1906), there is a median lacuna and pairs of lateral and intermediate lacunæ, agreeing in the main with the arrangement seen in the Glossiphonidæ. The hypodermal lacunæ are absent, but capillary vessels, having a similar function, occur in allied species. In the middle of each of the first eleven somites (XIII–XXIII) posterior to the clitellum there is a pair of lateral pulsating vesicles and a *segmental lacuna* running within the circular body in the form of an irregular ring.

The vesicles lie in the body-wall between the skin and the muscle-layers, and outside the lateral lacunæ, with which they are connected. The segmental lacunæ form a communication between the median and intermediate lacunæ, and throw out on either side lateral branches which, again, communicate with the vesicles.

The pulsating vesicles, which in diastole arch up the conspicuous little hemispheres of skin already referred to, are rendered contractile by reason of their muscular walls. An incomplete muscular septum, formed by an invagination of the outer wall, lies somewhat loosely within each vesicle, and by flapping backwards and forwards acts as a valve, alternately covering and uncovering the opening forming the vesicle's inlet. In expansion, the cœlomic fluid is drawn into the vesicle from the lateral branch of the segmental lacuna, and after passing over the septum is forced, during contraction, into the lateral lacuna.

Johansson (1898 *b*), working for the most part upon material obtained in northern Europe, draws attention to three different types assumed by the lacuna system in the Ichthyobdellidæ. The first type, possessing pulsating vesicles, is that just described which, it must be noted, contains one genus, *Pontobdella*, where these organs are not apparent externally, being very small and unable to arch up the thick warty skin. In the second type the lacuna system is considerably reduced. The pulsating vesicles and also the lower halves of the segmental lacunæ, have disappeared, leaving only the upper halves to connect the dorsal and lateral lacunæ. Johansson's genus *Abranchus* (unrecorded from India) and, perhaps, *Piscicola cœca* (Kaburaki), described in these

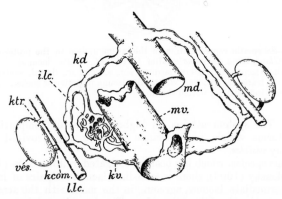

Fig. 4.—Schematic representation of the lacuna system in a somite in the middle part of the body of *Piscicola* (after Selensky). *md.*, dorsal lacuna; *mv.*, ventral lacuna; *l.lc.*, lateral lacuna; *i.lc.*, intermediate lacuna, shown on one side only; *kd.* and *kv.*, dorsal and ventral sections of segmental lacuna; *ves.*, pulsating vesicle; *ktr.*, canal connecting pulsating vesicle with lateral lacuna; *kcom.*, lateral canal running from segmental lacuna to vesicle.

pages, are representatives of this type. In the third type, exemplified by the genus *Platybdella* (Malm, 1863) as amended by Johansson (1898 *a*), the cœlomic system is reduced still further, little more than the ventral part of the median lacuna being left.

These types by no means include all the modifications exhibited by the lacuna system in the Ichthyobdellidæ, which still requires further investigation. Badham (1916) describes a remarkable Australian member of this Family, *Austrobdella*, having a pair of contractile marginal canals in place of pulsating vesicles.

In *Ozobranchus*, Oka (1904) finds that the contractile lateral lacunæ give off to each of the branchiæ a canal which breaks up into finer branches, of which two penetrate to the tip of every gill thread, where they unite with a similar single fine branch communicating with the large and simple median lacuna. In diastole the lateral lacunæ draw the lymph through the inlet valves of the branchial canals connected with them, and in systole disperse it through the rest of the cœlomic system.

VASCULAR SYSTEM.—The true Blood-Vascular system consists essentially, in the Rhynchobdellæ, of a dorsal and a ventral vessel extending through the greater part of the body, and connected at each extremity by a series of convoluted branches. A section of the anterior part of the dorsal vessel is provided with specially muscular walls, which render it contractile and so able to perform the functions of a heart or hearts. The dorsal vessel, furthermore, almost always expands into, or is connected with, what has been termed an *intestinal blood sinus,* which more or less entirely envelops the intestine and its diverticula.

In many, if not in all cases, the peristaltic contractions of the muscular intestinal wall force the blood out of the intestinal sinus into the dorsal vessel, where, after passing through a series of valves which prevent regurgitation, its forward flow receives fresh impetus on reaching the contractile " heart."

The vascular system here broadly outlined is subject to considerable variation in detail. The intestinal blood sinus in the Glossiphonidæ is described by Oka (1894) as an expansion of the dorsal vessel, entirely surrounding and following the contour of the intestine and its diverticula. In *Piscicola* and some other Ichthyobdellid genera this is formed in part by a separation of the epithelial and muscular layers of the intestinal wall. In *Branchellion* it takes very largely the form of a network of blood-vessels (Sukatschoff, 1912), and in *Ozobranchus* no proper sinus exists at all. In this genus, Oka (1904) finds that the dorsal vessel lies over the intestine, and the diverticula of the latter, which are very long and do not extend laterally, wind round the vessel so that it is in contact with the intestine on all sides. This close contact, which is always maintained between the dorsal vessel and the intestinal wall, whether by a sinus or by other means, enables the blood to absorb from the intestine food products which later are imparted osmotically through the walls of the dorsal vessel to the lymph surrounding it in the median lacuna.

Our knowledge of the vascular system in *Piscicola* is due to Johansson (1896) and Selensky (1907), and the latter gives a good description of the thickened muscular walls of the dorsal vessel forming the "heart" in this species (see fig. 5). These walls owe their contractility to the presence of closely-placed muscular bands surrounding the vessel, and extend from the second preclitellar somite to the posterior extremity of the clitellum, where they end abruptly. It will be observed that the muscular layer extends for a short distance along each of the branches given off by the dorsal vessel. In the Glossiphonidæ, Oka (1894) describes a somewhat different arrangement, consisting of a series of fifteen contractile chambers separated by valves, acting as hearts.

The anterior branches connecting the dorsal and ventral vessels consist nearly always of four pairs, together with an odd branch which supplies the proboscis. Posteriorly the connecting branches spread out into paired loops in the posterior sucker. In some cases there are seven pairs of such loops corresponding to the seven somites absorbed by the sucker, but the full number is not always present.

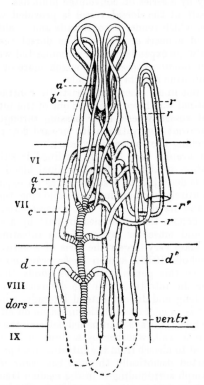

Fig. 5.—Schematic representation of the blood-vessels in the anterior part of the body of *Piscicola* (after Selensky). Somites numbered in Roman figures. *dors.*, dorsal vessel (the contractile portions are annulated); *ventr.*, ventral vessel; *a, a'* and *b, b'*, branch vessels forming loops in the anterior sucker and connecting the extremities of the dorsal and the ventral vessel; *c* and *d, d'*, other branches connecting the same, the latter looping backwards as far as the clitellum; *r, r'*, an unpaired connecting branch supplying the proboscis, here shown for the sake of clearness as external to the body.

It will be understood from what has been said that (1) in *Acanthobdella* there is a simple body-cavity and a closed vascular system, and that (2) in the Ichthyobdellidæ and Glossiphonidæ the closed vascular system has become supplemented by a secondary circulatory system developed from the cœlom. These two systems are unconnected and, moreover, the blood flowing in the one and the lymph flowing in the other, although both almost colourless, bear no other resemblance to each other and react differently to stains. In the Arhynchobdellæ, it may be noted further, the true vascular system has disappeared, and the lacuna system, containing red-coloured lymph, alone remains in a modified form.

ALIMENTARY TRACT.—The buccal orifice or mouth perforates the hollow ventral surface of the anterior sucker at some point in the middle line. The diagnostic value of the position of this orifice either within or upon the anterior rim of this *oral chamber*, as it has been called, has already been noted in the description of external features. The mouth opens into a deep stomodæal cavity completely surrounding the pharynx, which takes the form of a cylindrical, bluntly-pointed protrusible *proboscis* moving freely within it. This cavity has received more than one interpretation, and has been variously termed the buccal sinus, peripharyngeal chamber, pharyngeal sac and proboscis sheath. The *proboscis sheath*, as it will be called here, is lined by ectodermal epithelium, which is continuous with that of the proboscis and its lumen.

The proboscis, which is controlled by retractor muscles lying behind it and protractor muscles situated in its sheath, is itself highly muscular and extensile, and in addition to its longitudinal musculature, possesses a system of radial and circular muscles which, by expanding and contracting its lumen, provide a means of sucking blood. The ducts of the unicellular salivary glands, which lie outside the median lacuna and are often of very large size, enter the base of the proboscis and penetrate upwards to discharge their contents, some into its lumen and some at its extremity.

The ectodermal lining of the digestive tract ends with the base of the proboscis, which is immediately followed by the "anterior endodermal gut" of Sukatschoff (1912), lying in the clitellar region. This, which has sometimes been called the œsophagus, receives in many cases the ducts of a pair of lateral œsophageal glands. It may be distended owing to the development of paired lateral diverticula, or it may be somewhat long and slender, when it often undergoes a considerable amount of flexion during the retraction of the proboscis. This anterior endodermal gut is no more than an extension of the portion immediately following it, with which it is similar both in function and in structure, namely the long section of alimentary tract called by Sukatschoff the "anterior thin-walled part of the middle gut," and referred to in the following pages by its more familiar names of *crop* or *stomach*.

The chief function of the stomach is to act as a place of storage for food, and its capacity is found to vary inversely with the difficulty experienced by the leech in finding a host.

The stomach is provided posteriorly with a capacious extension, which in the Ichthyobdellidæ takes the form either (1) of a single cæcum, (2) of a pair of cæca, or (3) of a cæcum partly divided by a series of median apertures, representing a stage intermediate between the other two. Johansson (1898 b), who first drew attention to the diagnostic value of these varying structures, regards the paired cæca seen in his genus *Abranchus*, and in *Ozobranchus*, as the primitive type from which the others have been derived by a process of fusion. The single cæcum or blind gut lies beneath the intestine, and is characteristic of *Pontobdella* and the intermediate type, where the paired cæca are partly fused together, appears to be of the most frequent occurrence, and is seen in *Branchellion* and *Piscicola*. In addition to the posterior cæcum or cæca there is a series of anterior, metamerically disposed pairs of cæca or diverticula, the several pairs occupying the middle part of successive somites. In *Branchellion* and *Callobdella* the narrow portion of the gut lying between each pair of cæca is provided with an annular muscle-band forming a sphincter, and there is little doubt that such sphincters, dividing the stomach into a series of chambers, occur in similar situations throughout the Rhynchobdellæ.

In the Glossiphonidæ all the diverticula of the stomach are paired and often developed to a conspicuous degree, the deeper indentations in the irregular but symmetrical form often assumed being due to the interference of the dorso-ventral muscles.

The thin-walled crop or stomach opens through a sphincter into the *intestine*, the "thick-walled, glandular section of the middle-gut" of Sukatschoff (*loc. cit.*). This consists of a moderately wide, median, longitudinal tube, part of which may be ciliated, provided generally with four pairs of diverticula. It is here that the processes of digestion and absorption take place. The relation of this portion of the gut to the blood-stream has already been discussed.

The intestine opens, again through a sphincter, into the thin-walled *hind gut*, of which the posterior part serves as a *rectum* and discharges its contents through the dorsal and median *anus*, situated nearly always within the limits of somites XXVI and XXVII.

GENERATIVE ORGANS.—Leeches are hermaphrodite. The reproductive system in the Rhynchobdellæ is of great systematic importance, owing to the considerable diversity in detail often presented by its several parts. The male, like the female organs, are paired, with the exception of their common portions situated in the middle line.

The male organs may be considered first. The testes are disposed segmentally, and in the post-clitellar region lie between the diverticula of the crop or stomach. The number of pairs of testes varies in different genera. Their cavities form parts of the original cœlom, having arisen, like the ovaries, as proliferations of the epithelium of the lateral cœlomic cavities, and within them the process of spermato-genesis takes place. Each testis, or testicular sac, communicates by a short vas efferens with the vas deferens of its own side, and this, at some point not far in advance of the first pair of testes, leaves the tissues through which it has hitherto passed, and expands into an ejaculatory canal with muscular walls, often swollen and contorted, lying free in the ventral lacuna. Here the terminal portions of each canal turn inwards to form a median common part or prostate chamber, followed by an ectodermal invagination, referred to below, ending in the male orifice.

The posterior part of the ejaculatory canal, generally much coiled and separated from the terminal part by a constriction, constitutes an epididymis or sperm-reservoir where the spermatozoa, swept into it from the ciliated lumen of the vas deferens, become cemented together in compact bundles, and are stored for future use. The ejaculatory canals may be modified in several ways. The sperm-reservoirs may be reduced, or in certain cases absent. In *Glossiphonia heteroclita* (fig. 23, p. 61) these canals assume what may be regarded as their typical form. In *Glossiphonia complanata* (fig. 22, p. 59) the greater part of each canal is somewhat slender, and takes the form of a long and sinuous loop extending backwards in the median ventral lacuna as far as the twentieth somite.

The spermatozoa are packed in *spermatophores* for conveyance from one individual to another. These little structures, which have an outer chitinous envelope, consist usually of two more or less adherent, often club-shaped tubes containing sperm-bundles, tapering anteriorly to a blunt point and united below by a pedicel traversed by canals leading from each tube to the exterior. The foot, with its adhesive basal disc, is formed in the median, common, unpaired part of the male organs, or prostate chamber, and each of the paired terminal portions of the ejaculatory ducts, often called the prostate cornua, contributes one of the two tubes (Whitman, 1890; Brumpt, 1900). The spermatophore, when first produced, is nearly white, and may be seen readily by the naked eye. That of *Glossiphonia complanata* (fig. 6) ranges from about 5 to 8 mm. in length, but spermatophores of considerably smaller size are not infrequent and larger dimensions are sometimes attained.

The terminal part of the male organs usually takes the form of a small eversible bursa. In *Ozobranchus*, however, where the reproductive system presents features probably unique among the Hirudinea, Oka (1904) notes the occurrence of a true copulatory organ recalling the penis found in the Hirudidæ. The bursa, which is subject to a good deal of modification, is generally rudimentary in the Glossiphonidæ, consisting of little more than a small papilla pierced by the male pore. An exception to this is seen in *Theromyzon* (*Protoclepsis*) *tessellata*, where the bursa is fairly large and has been observed to transfer spermatophores of a reduced and simplified form from one individual to the female aperture of another (Brandes, 1900; Brumpt, 1900).

But although the well-developed bursa occasionally provides a means of copulation, fertilization, which may be reciprocal, is effected in the Rhynchobdellæ chiefly by hypodermic impregnation. This curious process consists (1) in the implantation by one leech of a spermatophore upon the body of another leech when (2) the contents of the spermatophore penetrate into the tissues of this other leech, and make their way to the ovarian sacs, where fertilization takes place. In the Glossiphonidæ the spermatophore probably proves effective if attached to almost any part of the

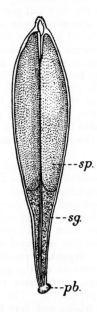

Fig. 6.—Spermatophore of *Glossiphonia complanata*, greatly enlarged (after Brumpt). *Sp.*, portion containing bundles of spermatozoa; *sg.*, granular secretion forming a temporary plug in the canal leading to the exterior through the pedicle *pb*.

body which happens to be accessible; its deposition, however, in many cases is made in the clitellar region. In most of the Ichthyobdellidæ, on the other hand, the spermatophore must be placed in a very restricted zone on the ventral surface, often differentiated into a *copulatory area* situated either close to the genital orifices or in some cases within the male atrium itself, when, occasionally, it may be brought to the exterior with the evaginated bursa. Beneath this area there lies generally the so-called *conductive tissue* (tissu vecteur) of Brumpt, serving as a passage through which the spermatozoa travel to their destination.

The female reproductive organs consist of a pair of more or less elongated cœlomic sacs—the ovisacs—and these, which contain the true ovaries, unite anteriorly to form a common muscular oviduct opening to the exterior by the female pore. The ovisacs during the breeding-season increase in length, and become much distended by the egg-strings developing within them, and in the case of the Glossiphonidæ and certain Ichthyobdellidæ lie free in the ventral lacuna.

In many Ichthyobdellidæ, however, the ovisacs form adherences either directly, or through the medium of conductive tissue, with

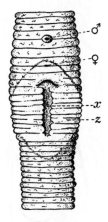

Fig. 7.—Ventral surface of clitellar region of *Piscicola geometra*, showing copulatory area *z*, provided with a longitudinal groove *x* (after Brumpt).

parts of the body-wall often marked exteriorly by a copulatory area of the kind to which reference has been made. This conductive tissue, which has sometimes been misinterpreted, varies very greatly in development and distribution, appearing in some cases as a fairly compact mass, and in others being reduced to a pair of bridle-like strands (fig. 8). It consists of a form of connective tissue arising as an outgrowth from the walls of the ovisacs, according to Brumpt (1901), who first drew attention to its special function, and to whose comprehensive work the reader is referred. The exceptional character of the male reproductive organs of *Ozobranchus* has already been mentioned. The female organs described by Oka (1904) in this genus are even more peculiar. Here there are two female apertures, that serving for fertilization being situated within the male atrial chamber and having no connection with the external female pore reserved for the deposition of eggs.

The eggs, in *Ozobranchus*, are embedded in a single layer in a chitinous sheet spread upon the plastron of the tortoise serving as a host (Plate II, fig. 5), but in most of the Ichthyobdellidæ they are enclosed in cocoons or egg-cases, which are attached to some submerged foreign body. These cocoons vary a good deal in form and size, but they are all made upon the same general plan.

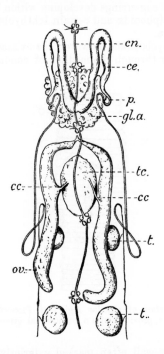

Fig. 8.—Reproductive organs of *Piscicola geometra*, showing conductive tissue (after Brumpt). *t.*, testis; *ce.*, ejaculatory canal; *p.*, its terminal portion; *gl.a.*, glands communicating with the terminal portion of the ejaculatory canal and secreting the walls of the spermatophore (these glands in most species lie in the walls of the terminal portion); *tc.*, mass of conductive tissue underlying the copulatory area; *cc.*, its paired connections with the ovisacs *ov.* *cn.*, ventral nerve cord with ganglia.

Shortly before a leech is ready to lay, it makes itself fast by its two suckers to some convenient object, and the clitellum becomes covered with a chitinous layer secreted by glands situated within its walls and destined to form the future cocoon. At first this layer is nearly white and somewhat viscous, but it gradually hardens and eventually assumes a more or less deep brown colour. The leech has now to withdraw the anterior part of its body through this cylindrical layer, having first discharged its eggs within it and caused it, by pressure, to adhere to the object upon which it rests. Swelling itself against the posterior extremity of its chitinous belt and dragging in the anterior extremity with it as it retires, this structure becomes invaginated until its two ends meet, thus preserving its contents from contamination by contact with the body passing through it. As soon as the cocoon is left free it evaginates itself, and its extremities close up, leaving a small operculum through which, when hatched, the young leeches issue into the surrounding water.

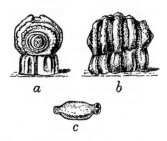

Fig. 9.—*a* end view, and *b* side view of cocoon of *Pontobdella muricata*, greatly enlarged; *c*, cocoon of *Piscicola geometra*, enlarged (after Harding).

In the Glossiphonidæ the typical egg-case is made in much the same way as the Ichthyobdellid cocoon. It differs from the latter, however, (1) in being a very thin and transparent membranous sac, and not a horny structure, and (2) in being attached to the ventral surface of the body from which it arose. A leech bearing more than one of these bulky and easily detachable sacs remains, as far as possible, in one place, a circumstance which has led to the supposition that the Glossiphonidæ brood over their eggs. The young leeches for some time after they are hatched adhere by their posterior suckers in heavy clusters to the ventral surface of the parent, who often directs the sides of her body downwards in order to afford additional protection and carries her brood to their first host. Instances are known of a dorsal gland secreting an adhesive substance whereby the young leech in its early stages is fixed to its parent. The dorsal chitinous plate conspicuous in *Helobdella stagnalis* is the remnant of such an attachment gland.

We have now given a brief sketch of the more important features, both external and internal, of systematic importance in the Rhynchobdellæ, without attempting, however, to enter into histological detail or to deal comprehensively with the morphology of the group. Johansson (1898 *b*) first called the attention of the systematist to the nephridia, and their diagnostic value, particularly in the case of the Ichthyobdellidæ, is becoming better appreciated. Our knowledge of many known species of Rhynchobdellæ, it may be observed, leaves much to be desired, and there can be little doubt that further species await discovery and investigation.

METHODS OF PRESERVATION AND STUDY.—The remarks upon the proper technique to be employed in the preservation and study of the Arhynchobdellæ made by Prof. Moore in this volume (pp. 117–120) apply equally well to the sub-order under consideration. Leeches are particularly difficult to preserve with their diagnostic features intact, and the greatest care should be exercised in their treatment after capture. Too often the well-meaning collector drops these sensitive worms alive into a preservative fluid frequently of unknown strength, with the result that they reach their destination in a condition which defies examination. The collector should endeavour to preserve his captives alive until he has leisure to examine them in their natural element, if necessary with the aid of a lens. Careful notes should then be made upon coloration, pattern, form, size, behaviour and any other points which attract the attention of the observer. Such notes, forwarded with the material concerned, often prove of great value.

Leeches invariably should be anæsthetized before being placed in the preservative fluid. This is most readily effected in all but the larger Rhynchobdellæ by immersion in water impregnated with carbon dioxide, obtainable nearly everywhere in the form of the beverage, soda-water. For the amateur 70 per cent. alcohol is the safest preservative to use; where possible, however, both alcohol or the still better preservative formalin should be applied as directed by Professor Moore.

In conclusion, I am greatly indebted to the late Dr. Annandale sometime Director of the Indian Zoological Survey, not only for the large amount of material placed in my hands, but also for valuable information upon the coloration and habits of many species, often supplemented by water-colour drawings of the living leech made by Mr. Chowdhary of the Indian Museum. I have received useful material and information from Miss Muriel Robertson, Prof. Clifford Dobell, F.R.S., Dr. Guy A. K. Marshall, C.M.G., Dr. Kaburaki, and also from Prof. Percy Moore, who, in addition to other courtesies, assisted me in connexion with *Placobdella ceylanica*, a species which owes its proper determination more to him than to me. My thanks are also due to Prof. G. H. F. Nuttall, F.R.S., for permission to reproduce from ' Parasitology ' drawings of the cocoons of *Pontobdella* and *Piscicola* (p. 29) and of *Helobdella stagnalis* (fig. 27, p. 69); and also to Major R. B. Seymour Sewell, I.M.S., the present Director of the Zoological Survey of India, who has added materially to the value of these pages by lending the blocks of the various figures here reprinted from the ' Memoirs ' and ' Records ' of the Indian

Museum. These acknowledgments would be incomplete without reference to the courtesy of Prof. J. Stanley Gardiner, F.R.S., who allowed me to make full use of the Cambridge Zoological Laboratory, where much of this work has been carried out; and also to the kind assistance of Sir Arthur Shipley, G.B.E., F.R.S., the editor of this volume, to whom my special thanks are due for much valuable criticism and advice.

Cambridge, May 17th, 1926.

CONTENTS.

D

HIRUDINEA.

Vermiform hermaphrodite Chætopoda provided with a sucker at both extremities, with median genital openings, without parapodia and rarely with branchiæ. With the exception of *Acanthobdella*, there are no setæ and the cœlom is broken up into a system of inter-communicating spaces.

The systematic position of *Acanthobdella* has been referred to in the Introduction.

The Hirudinea are divided into the two suborders, Rhynchobdellæ and Arhynchobdellæ.

Suborder RHYNCHOBDELLÆ.

Marine and freshwater Hirudinea with colourless blood, with an exsertile proboscis, without jaws. The mouth is a small median aperture situated within the anterior sucker, or rarely upon its anterior rim.

Family ICHTHYOBDELLIDÆ.

Body cylindrical or flattened, often divided into two distinct anterior and posterior regions, and sometimes with paired lateral branchiæ and pulsating vesicles. The anterior sucker is generally, and the posterior sucker is always, a permanent cupuliform or discoid organ distinct from the body. Eggs either included in chitinous capsules, which are attached to foreign objects, or cemented to the body of the host. Marine and freshwater forms, largely parasitic upon fish.

In the case of Ichthyobdellidæ having the body divided into two regions, the anterior region includes the clitellum with the genital orifices, and the whole or part of somite XIII always forms the beginning of the posterior region. The anterior portion is conveniently termed the neck, and the posterior region is, similarly, called the trunk or abdomen. The somewhat heterogeneous assemblage of leeches contained in this Family obviously requires subdivision, and various attempts to split it up have already been made. That none of these attempts has proved entirely satisfactory gives no cause for surprise, since it is now generally accepted that external features alone cannot often be relied upon for purposes of classification, and our knowledge of the internal morphology of the Ichthyobdellidæ leaves much to be desired. This knowledge necessarily must be slowly acquired, for the members of this Family are notoriously difficult to preserve in a satisfactory state, and the marine species, usually found in unexpected places and under unfavourable conditions, rarely fall alive into competent hands.

Genus OZOBRANCHUS, de Quatrefages, 1852.

Ozobranchus, de Quatrefages, 1852, p. 325.
Eubranchella, Baird, 1869, p. 311.
(?) *Lophobdella*, Poirier et de Rochebrune, 1884, p. 1597.
Pseudobranchellion, Apáthy, 1890, pp. 110 and 122.

Marine and freshwater Rhynchobdellæ parasitic for the most part upon turtles and tortoises. Body more or less flattened and divided into two distinct regions, a short, narrow, anterior "neck" and a large broad posterior portion or "abdomen." Posterior region with paired, lateral digitate branchiæ; without pulsating vesicles. One pair of eyes. The complete somite is bianuulate anteriorly, but may become triannulate in the posterior region. Eggs cemented to the body of the host. Oral opening subterminal.

1. Ozobranchus shipleyi, Harding, 1909. (Plate II, figs. 4, 5, 6; and fig. 10.)

Ozobranchus shipleyi, Harding, 1909, p. 233.
Ozobranchus jantseanus, Kaburaki, 1921 *b*, p. 681 (not Oka, 1912).
Ozobranchus papillatus, Kaburaki, 1921 *b*, p. 692.

The first brief description which I gave of this species was based upon some small individuals taken from the terrapin, *Nicoria trijuga*, in Ceylon by Miss Robertson (*q. v.* 1910). These were not in a favourable condition for the examination of external features, and I stated at the time that the complete somite was formed of three rings. A careful re-examination of the original leeches and their comparison with fresh material has convinced me that the complete somite is really biannulate, consisting, in the neck, of two rings of almost equal size, and in the posterior region of a broad ring followed by a narrow one. The broad ring, however, on the abdominal dorsal surface often shows signs of transverse subdivision, but this is not carried far enough to resolve it into two distinct rings and so constitute a true triannulate somite.

Description.—Body flattened and translucent; the posterior region with eleven pairs of lateral digitate branchiæ, a pair occurring on the anterior ring in each of the somites XIII–XXIII.

The mouth opens in the anterior sucker in a subterminal position. Posterior sucker large, circular, centrally attached and about equal in width to the broadest part of the body (excluding the branchiæ).

According to the late Dr. Annandale, to whom I am indebted for

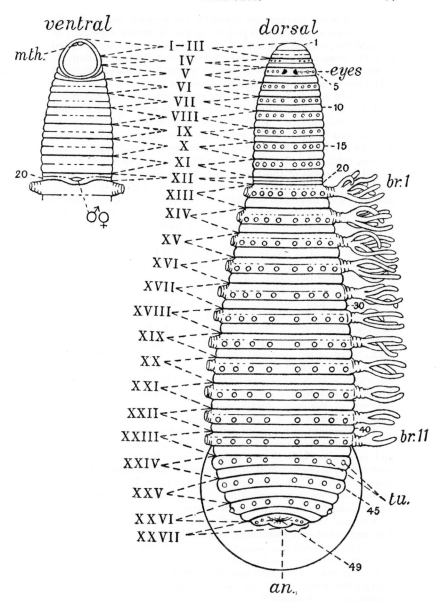

Fig. 10.—*Ozobranchus shipleyi*, Harding, 1909. Diagram showing dorsal and part of the ventral surface. Somites numbered in Roman, and rings in ordinary figures. *mth.* Mouth. *an.* Anus. *br.* 1, *br.* 11. First and eleventh pairs of branchiæ. (The branchiæ are shown fully on one side only, for the sake of clearness.) *tu.* Tubercles.

notes on the appearance and habits of this species during life, the whole dorsal surface is dull yellow, delicately varied, especially at the margins, with dark green. Posterior sucker minutely speckled with the same green; branchiæ colourless and almost transparent.

Rings 49. The second and the forty-ninth rings show signs of subdivision, but are here regarded as single rings.

Complete somite formed of two rings, the anterior one being distinguished by a series of 10–12 papillæ of varying size situated on the dorsal surface. In the "neck" region these two rings are more or less of equal width, but in the posterior region, beginning with somite XIII, the anterior ring is conspicuously larger than the one behind it. The branchiæ, when present, spring from the lateral extremities of this broad ring, which, as in the case of Oka's leech, *Ozobranchus jantseanus*, represents rings 1+2 of a triannulate somite. This inference is emphasised by the fact that the papillæ present upon this ring occupy its posterior part, and are often, but not always, separated from its anterior part by a shallow transverse groove, as noted above.

Somites I–III are represented by the first two rings.

The twenty-three somites IV–XXVI are complete with two rings. Somite XXVII is uniannulate.

The single pair of eyes lie in ring 5 (the first ring of somite V).

The male and female genital ducts open by a common pore between the two rings (19 and 20) of somite XII. The reproductive organs are of the complicated type characteristic of the genus. There are four pairs of testes, occupying somites XVI–XIX. The small, spherical but somewhat flattened eggs have a tough shell provided with a minute circular aperture closed by a lid, which is broken through by the young leech when it emerges. These are laid close together in large groups on the carapace or plastron of the host, partly embedded in a layer of chitinous cement, which may be stripped in small sheets from the host's body. The anus opens in the anterior part of the last ring.

The long tubular crop, or stomach, increases in size posteriorly, and in somite XIX throws out two large lateral cæca, directed posteriorly, between which lie the intestine and rectum. The four pairs of intestinal cæca are not spread out laterally, but lie closely packed together in a longitudinal bundle, confined, as Oka (1904) has explained, within the median lacuna.

A typical branchia consists of a short, thick, circular, more or less annulated and sometimes bluntly-branched stalk with long vermiform unbranched appendages issuing from its distal end. The branchiæ are kept in constant slow motion when the living leech is at rest "as though the thumb and fingers of a hand were continually being slowly opposed to one another and as slowly withdrawn" (*Annandale*).

Dimensions.—Length, in alcohol, up to about 25 mm.; greatest width, not including the branchiæ, about 5 mm. Larger dimensions probably are attained. The Indian individuals were generally much larger than those from Ceylon, but this seemed insufficient evidence for regarding them as a separate species.

Hosts and *Habitat.*—This species has been found on *Nicoria trijuga*, Schweigg, in Ceylon. Indian examples have been taken from *Kachuga intermedia*, Blanf., collected in the R. Mahanaddi, Sambalpur, Orissa; from *K. smithii*, Gray, in the R. Ravi, Lahore; and from a single specimen of *K. dhongoka*, Gray. The last-mentioned tortoise originally came from the R. Ganges, but was in captivity in the Zoological Gardens, Calcutta, when the leeches were taken from it. They were adhering tightly, in a solid mass, to the plastron, and were difficult to remove.

Genus **PONTOBDELLA**, Leach, 1815.

Albione, Savigny, 1822.

Marine leeches without eyes, pulsating vesicles or branchiæ. Body more or less claviform and covered by conspicuous tubercles, often very large and forming prominent warty protuberances. Crop (or stomach) with a single undivided cæcum. Complete somite composed of three and sometimes of four rings. Parasitic on fish, and chiefly affecting skarks, skates and rays.

Of the three species here described, two have the complete somite composed of three rings; in the third species, however, this is composed of four rings. Two species, again, have the typical claviform outline; *P. macrothela*, on the other hand, differs from them entirely in the form of the body. Although the number of rings of which the complete somite is composed is of great diagnostic importance, the genus *Pontobdella* is here regarded, provisionally at least, as being unique in possessing a variable complete somite. It may, perhaps, become necessary at some future time to re-arrange the various species now referred to this genus, when a clearer knowledge of their internal morphology has been obtained. It was not deemed advisable, however, to procure such knowledge, in the case of the three species dealt with here, by sacrificing the scanty and valuable material available. In any case, the striking difference in body-form shown by *P. macrothela* seems sufficient to justify its inclusion in a subgenus, which has here been named *Pontobdellina*.

2. **Pontobdella loricata**, Harding, 1924. (Plate II, fig. 3; and fig. 11.)

Description.—Body fusiform and of a uniform greyish-green colour in alcohol, unrelieved by special markings. Anterior sucker circular, cupuliform, excentrically attached, with three pairs of inconspicuous submarginal papillæ and with corrugated edges capable of being folded together so as to form a ventral, median, longitudinal slit. It comprises the first four somites and the greater part of the fifth.

Posterior sucker circular, centrally attached, with corrugated edges, not wider than the anterior sucker or the greatest width of the body.

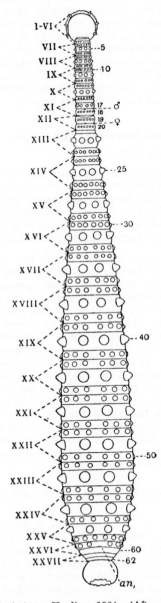

Fig. 11.—*Pontobdella loricata*, Harding, 1924. (After Harding.) Diagram
showing external features of the dorsal surface. Somites numbered in
Roman, and rings in ordinary figures. *an.* Anus.

Rings, 62 behind the anterior sucker. Complete somite formed of three annuli, consisting of a broad middle ring (lodging a ganglion of the ventral chain) situated between two smaller rings of equal width.

Somites VII–XI and XIII–XXIV complete with three rings; XII, XXV, XXVI and XXVII biannulate.

The conspicuous clitellum comprises the seven rings lying between the broad middle annuli of somites X and XIII. Portions of a ring may be detected between the 18th and 19th annuli. This appears to represent the first annulus of somite XII (here regarded as missing), but should an examination of further material prove it to be a constant feature, it might well be included in an enumeration of the annuli, thus bringing the total number of rings to 63 and rendering somite XII triannulate.

The warts or tubercles upon the rings vary a good deal in number, size and position, so that it is only possible, by striking an average, to arrive at what may be considered to be a normal arrangement. This normal arrangement, shown, as far as the dorsal surface is concerned, in fig. 11, may be described as follows :—

The broad middle ring of the somite bears on its whole circumference eight large conical tubercles, four above and four below. The exterior dorsal warts are submarginal, so that their points appear in a ventral view. The narrow rings on either side of the middle one each bear a total of fourteen tubercles, a pair of marginal tubercles being present in addition to six above and six below.

The warts here described may be occasionally missing or out of their normal position, and other tubercles, generally of small size, may be interposed between them. On the ventral surface of the middle ring of the somite a small tubercle often occurs between the middle pair of warts. The mouth-opening occupies a nearly central position within the anterior sucker.

The male genital orifice lies between rings 17 and 18, that is, between the second and third rings of somite XI; the female opening appears between rings 19 and 20, in the middle of somite XII. Crop (or stomach) with a sacculated, single, undivided cæcum.

The anus opens between rings 59 and 60, in the middle of somite XXVI, and so is separated by three annuli from the posterior sucker.

Dimensions.—Approximate total length, in alcohol, 74 mm.; approximate greatest width 7 mm.

Host and *Habitat.*—The host is not recorded, and the example here described is noted as having been taken at station 233, Marine Survey of India.

3. Pontobdella aculeata, Harding, 1924. (Figs. 12 & 13.)

Description.—Body fusiform, much attenuated anteriorly, more or less circular in transverse section. One of the two individuals examined, in alcohol, had assumed a uniform dull grey hue devoid

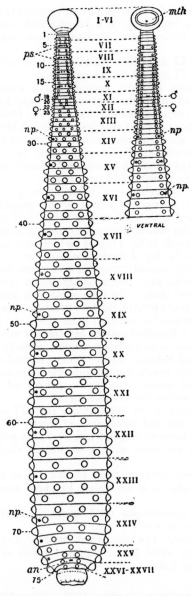

Fig. 12.—*Pontobdella aculeata*, Harding, 1924.   (After Harding.)   Diagram
showing external features on the left of the dorsal, and on the right
of part of the ventral surface.   Somites numbered in Roman, and rings
in ordinary figures.   *p.s.* Pigment spots.   *mth.* Mouth-opening.   *an.*
Anus.   *np.* Nephridiopores (on ventral surface).

of special markings; the other was of a reddish-brown colour with a dorsal pair of dark brown linear spots on the sensory ring of each somite. These spots appeared to recur metamerically throughout the body, but were difficult to detect owing to the imperfect preservation of the material.

Anterior sucker small, circular, without papillæ, having corrugated edges tending to close together in a transverse line and absorbing somites I–IV and the whole, or the greater part, of somite V.

Posterior sucker smaller than the anterior; with corrugated edges, circular and centrally attached.

Rings, 75 behind the anterior sucker. Complete somite formed of four annuli of nearly equal width. The third of these, conspicuous on account of its larger marginal papillæ, is here regarded as the primary ring of the somite, although it does not entirely

Fig. 13.—*Pontobdella aculeata*, Harding, 1924. Dorsal aspect, life size.

cover a ganglion of the ventral chain. The ventral ganglia, throughout the greater part of the body, lie almost exactly between this third ring and the one anterior to it. The ring in question, however, takes its proper position in the middle of the triannulate somites present.

Somites IX, X and XIII–XXIV complete with four annuli; VII, VIII, XI and XII triannulate; XXV biannulate; XXVI and XXVII uniannulate.

The clitellum is not conspicuous, and appears to comprise the seven rings 17–23. The arrangement of the warty tubercles on the body, shown in fig. 12, may be summarized as follows:—

(1) Each annulus bears two lateral tubercles, which, as already stated, are larger on the primary than on the other annuli of the

somite. These tubercles form two conspicuous rows lying along the margins of the body.

(2) Dorsally, the first and third rings of the complete somite bear a paramedian pair, and the second and fourth rings a paramarginal pair of tubercles together with a median one, so that, throughout the greater part of the body, the tubercles on the upper surface are disposed alternately in twos and threes, forming a symmetrical pattern.

(3) On the otherwise bare ventral surface of each ring is a paramarginal pair of prominent tubercles. These tubercles constitute two conspicuous ventral rows, which form a characteristic feature of the species.

The mouth opens in a nearly central position within the anterior sucker.

The male genital pore is situated between rings 19 and 20, the second and third rings respectively of somite XI: the female opening lies between rings 22 and 23, that is, between the second and third rings of somite XII.

The crop has a single undivided cæcum, and the anus perforates the seventy-fourth or penultimate ring. The nephridiopores open in a paramarginal position in the second ring of the complete somite. Thirteen pairs of these pores were observed, but the examination of fresh material will be necessary before their full number can be determined.

Dimensions.—Approximate size, in alcohol, of the larger example, total length 64 mm., greatest width 8 mm.; of the smaller, total length 35 mm., greatest width (gorged with blood) 6 mm.

Hosts and *Habitat.*—(*a*) Found on *Harpodon nehereus,* Estuary of the Bassein River, Burma (Marine Survey of India). (*b*) From the Gregory Isles, Mergui Archipelago, Burma (Marine Survey of India).

The teleostean fish *Harpodon nehereus,* Ham. Buch., when properly salted and dried, forms the table delicacy known as Bombay-duck.... " It is not known to occur at any great depth, and is not even restricted to the sea, being very abundant in the rivers and estuaries of Bengal and Burma " (G. A. Boulenger, 1904, p. 613).

PONTOBDELLINA, subgen. nov.

Marine leeches having the characters of *Pontobdella,* with the exception of the form of the body, which is sharply divided into two regions—a slender anterior " neck " and a broad, posterior " abdominal " region.

See note on the genus *Pontobdella.*

4. Pontobdella (subgen. Pontobdellina) macrothela, Schmarda, 1861. (Plate II, fig. 8 ; and figs. 14 & 15.)

(?) *Hirudo indica*, Linnæus, 1767, p. 1079.
(?) *Albione indica*, Moquin-Tandon, 1826, p. 130.
(?) *Pontobdella indica*, de Blainville, 1827, p. 243.
(?) *Pontobdella depressa*, Krøyer (see Diesing), 1850, p. 438.
Pontobdella macrothela, Schmarda, 1861, p. 6, Taf. xvi; R. Blanchard, 1897, p. 80 ; Goddard, 1909, p. 721.

The *Pontobdella depressa* of Krøyer appears to be synonymous with *P. macrothela*, but its description is insufficient to place its identity beyond dispute. According to the communication made to Diesing by Krøyer, his leech was found in West Indian waters, and had an ashy-yellow, flattened body with flattened and deeply furrowed warts. Regarding the *Hirudo indica* of Linnæus, another species of *Pontobdella* described as being flattened

Fig. 14.—*Pontobdella* (subgen. *Pontobdellina*) *macrothela*, Schmarda, 1861. Dorsal aspect, life size.

(*depressa*), we can do no more than remark that it seems to bear a greater resemblance to *P. macrothela* than to the other Indian species of *Pontobdella* so far recorded.

Description.—Body much flattened and divided into two very distinct regions : a short slender "neck" and a long, bulky "abdominal" region comprising somites XIII–XXVII. Dorsal surface with a longitudinal median groove, conspicuous on the posterior region.

The colour varies from yellowish-brown to dark green. The example from India was, according to the collector's note, in life,

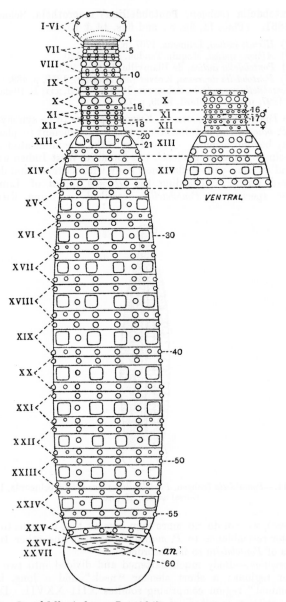

Fig. 15.—*Pontobdella* (subgen. *Pontobdellina*) *macrothela*, Schmarda, 1861. Diagram showing external features on the left of the dorsal and on the right of part of the ventral surface. Somites numbered in Roman, and rings in ordinary figures. *an*. Anus.

dark green. This had faded in alcohol to a pale greenish-brown hue, and the body was unicolorous, save for two dorsal, dark brown spots on the anterior sucker above the junction with the " neck," symmetrically placed one on either side of the median line. Both Schmarda and Goddard state the colour to be yellowish-brown : the former gives no indication of body-markings in text or figure, and the latter definitely notes their absence.

Anterior sucker slightly oval, having its long axis parallel to the annuli of the body; about half the width of the broadest part of the body; with six small, conical, submarginal, dorsal tubercles, and showing signs of annulation dorsally above the junction with the " neck." It appears to comprise the first five somites.

Posterior sucker large and oval, with its long axis lying along the middle line; about equal in width to the broadest part of the body.

Annuli 60; the first annulus to encircle the body completely, being the anterior ring of somite VI. Rings 58, 59 and 60 without warts, but deeply scored by irregular transverse furrows tending to render obscure the interannular grooves. Complete somite formed of three annuli, a broad one situated between two narrow ones. In typical somites the narrow annuli are of equal width and about half the width of the middle one.

Somites VII–X and XIII–XXIV are complete with three rings. Somites XI and XII (in the clitellum) with XXV and XXVI bi-annulate; XXVII uniannulate.

Throughout the greater part of the body the ventral ganglia lie in the broad middle ring of each somite.

The prominent, irregularly-shaped warty tubercles which cover the whole body have a rough, nodular surface, and are often of extensive area. In the "abdominal" region, where they tend to expand into and fill up the spaces between each other, they are separated by deep furrows, and, together with a series of large depressed warts of more or less quadrangular outline situated on the middle ring of each somite, present a striking appearance as of crocodile skin (Pl. II, fig. 8.)

The warts on the middle ring of the complete somite are twelve in all and arranged as follows : dorsally four large, more or less quadrangular warts, two on either side of the middle line, and between each of these two a small wart; ventrally six warts, the two lying on either side of the middle line being smaller and less conspicuously quadrangular than the others.

In the two narrow annuli of the somite, Goddard makes the number of warts, or "tubercular areas" as he calls them, the same, viz. eight above and eight somewhat smaller ones below. These observations apply to the Indian example here described, with the exception of the number of dorsal warts on the first annulus of the complete somite, where only six were usually present. No doubt some variation occurs both·in their number and position.

Mouth nearly centrally placed within the anterior sucker. Male genital orfiice situated between annuli 16 and 17, in the middle of somite XI; female genital orifice situated two rings behind the male, between annuli 18 and 19, in the middle of somite XII.

The anus lies between annuli 57 and 58, that is, between somites XXV and XXVI.

Dimensions.—Size of Indian specimen examined (fairly extended, in alcohol): length, 64 mm.; greatest width, 13 mm.; anterior sucker, fully expanded, 6 × 5 mm.; posterior sucker, 13 × 11 mm.; length from beginning of "trunk" region (somite XIII) to extremity of posterior sucker, 55 mm.

Greater dimensions are recorded. Length (alive, but quiescent), 95 mm.; greatest width, 20 mm. (*Schmarda*): an example in alcohol, 80 mm. in length, 10 mm. in breadth (*Goddard*).

Habitat and *Hosts.*—The description here given is based on a single example of this species found, according to information supplied by its collector, attached to the side of a hammer-headed shark (*Zygæna* sp.) caught at a depth of 25–28 fathoms off Gobalpore, in the Bay of Bengal.

Blanchard records this leech from Tandjong, Lampongsche Districten, Sumatra; Goddard describes and figures an individual from the Brisbane River, Australia; and Schmarda founded the species on a solitary example taken in the harbour of Kingston, Jamaica.

The only host of *P. macrothela* yet recorded is the hammer-headed shark just referred to. The appearance of this leech in parts of the world so far apart as Kingston Harbour, the coast of New South Wales and the Bay of Bengal is consistent with parasitism upon certain sharks of wide distribution such as *Zygæna malleus* and *Z. tudes*, and its habitat probably includes sub-tropical as well as tropical seas.

Genus **PISCICOLA**, de Blainville, 1818.

Small freshwater and brackish-water leeches parasitic generally upon fish. Body much attenuated, smooth and cylindrical, the posterior region with paired, lateral pulsating vesicles. Suckers large and excentrically attached. Mouth-opening in the middle of the anterior sucker. Four eyes, generally linear in form, upon the anterior sucker. Complete somite composed of fourteen rings.

5. **Piscicola olivacea**, Harding, 1920. (Figs. 16 & 17.)

Piscicola olivacea, Harding, 1920, p. 512; Kaburaki, 1921, p. 663.

Description.—Body circular, long and slender, varying in colour from bright to pale olive-green, minutely speckled with black or with a deeper shade of green.. A series of conspicuous white patches or spots occur, one on either side of each somite, on the margins of the body, and these are connected across the dorsal surface by whitish, often indistinct transverse bands. Another series of white patches of more or less elliptical form lie in the mid-dorsal line, and these median patches, which may or may not be jointed together at their extremities, give the appearance of a somewhat ill-defined whitish mid-dorsal stripe.

Anterior sucker circular, whitish, with three brown dorsal transverse bands, one band near the anterior extremity, one following the junction with the body, and a third and broader one between the two, in the posterior part of the sucker. The mouth-opening lies in the centre of its interior cup.

Posterior sucker somewhat heart shaped, of the same green colour as the body, with seven pairs of whitish rays corresponding to the seven somites XXVIII–XXXIV, of which it is composed.

Eyes, two pairs, lying within the middle brown band on the anterior sucker, one pair on either side of the mid-dorsal line. The component eyes of each pair are linear in form and, without actually touching, are inclined together at an acute angle having its vertex pointing towards the margin of the sucker.

Complete somite formed of 14 rings. An accurate count of all the rings present in this little species could not be made in the material examined.

A pair of lateral pulsating vesicles occur in each of the eleven somites XIII–XXIII. These vesicles are centred in the transverse middle line of the somite and lie within the marginal white spots.

In the middle part of the body a ventral ganglion occupies rings 7 and 8 of the complete somite.

The anus is situated in the middle of the last ring.

The male genital opening lies in the posterior part of somite XI, and the female opening in the posterior part of somite XII.

The external features of this species are shown in fig. 16, and it is to be understood that the dorsal pattern, which is subject to some variation in its details, is merely indicated schematically. The internal investigation of *P. olivacea* has been undertaken by Dr. Kaburaki, to whom I am indebted for the information conveyed by fig. 17. There are six pairs of testes. The intestine leaves the crop (or stomach) at a point behind the sixth pair of lateral diverticula, between somites XVIII and XIX, and the posterior portion of the crop, consisting of a single, undivided cæcum, extends beneath it as far as somite XXV and is also provided with lateral pouches.

Dimensions.—Approximate total length when fairly extended, 10·75 mm., the greatest width of the body being about 1·50 mm.

Hosts and *Habitat.*—*Picicola olivacea* has so far only been recorded in India from the Chilka Lake, where it is of frequent occurrence in more or less brackish water. It has been taken there :—

(1) From the Sting Ray, *Trygon sephen*, Forsk. "On the lower surface of the body, in the gill-slits, near the anus

E

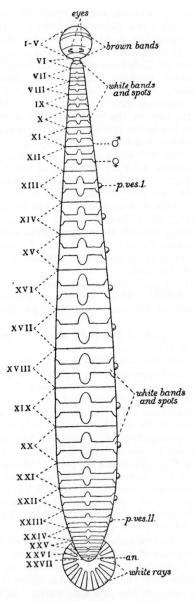

Fig. 16.—*Piscicola olivacea*, Harding, 1920. Diagram showing dorsal pattern and other external features. Somites numbered in Roman figures. *p.ves.* 1, *p.ves.* 11. Pulsating vesicles (shown on one side only). *an.* Anus. (After Harding.)

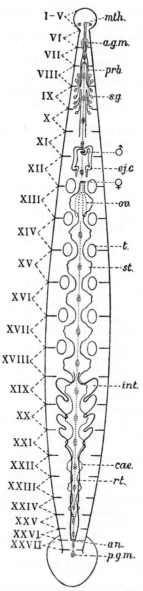

Fig. 17.—*Piscicola olivacea.* (After Kaburaki.) Diagram showing alimentary
tract, reproductive system and ventral nerve-cord. Somites numbered
in Roman figures. *a.g.m.* Anterior ganglionic mass. *p.g.m.* Posterior
ganglionic mass. *gang.* Ventral ganglion. *mth.* Mouth-opening. *s.g.*
Salivary glands. *st.* Stomach. *int.* Intestine. *rt.* Rectum. *cæ.* Cæcum.
an. Anus. *ej.c.* Ejaculatory canal. *ov.* Ovary. *t.* Testis.

and within the mouth on the palate." (sp. gravity of water 1·007–1·011).

(2) From the Globe fish, *Tetrodon reticularis*, Bl. (sp. gravity of water about 1·026).

(3) From *Chatoessus chacunda*, Ham. Buch.

The late Dr. Annandale recorded specimens from a small pool of almost fresh water on Barkuda Island, Lake Chilka, and noted that in this pool the only vertebrates were frogs (*Rana cyanophlyctis*).

P. olivacea also occurs in China, two specimens from Soochow having been described by Moore (1924, p. 346).

6. Piscicola cæca, Kaburaki, 1921. (Fig. 18.)

This species can only be referred to the genus *Piscicola* provisionally. Although possessing certain features seen in *Piscicola*, the absence of pulsating vesicles, with the important modification of the cœlomic system which this absence implies, together with the absence of eyes, is sufficient to distinguish it from members of that genus. It would be unwise to dogmatise further, however, without additional investigation and until certain Ichthyobdellid genera which are still subjects of controversy are more clearly defined. The following brief diagnosis of *P. cæca* (of which no example has been seen by the writer) is based on the detailed description given by Kaburaki (1921, p. 666).

Description.—The slender, fusiform, translucent body is much flattened and has faded in alcohol to a uniform greyish-white colour without any trace of pattern.

Anterior sucker nearly circular, cup-shaped and about half as wide as the heart-shaped posterior sucker.

No eyes and no lateral pulsating vesicles.

Complete somite formed of 14 rings.

The male genital opening lies in somite XI, and the female opening in somite XII, fourteen rings behind the male. There are six pairs of testes.

The digestive tract closely resembles that of *Piscicola olivacea*. The mouth opens in the middle of the cupuliform anterior sucker, and the anus lies between somites XXVI and XXVII. The crop (or stomach) is provided with ten pairs of lateral diverticula. Six pairs of these diverticula lie in front of the junction with the intestine (between somites XVIII and XIX) and the remaining four occur in the single, undivided cæcum which extends beneath it.

Dimensions.—Length, 13 mm, ; greatest width, approximately, 1 mm.

Hosts and *Habitat.*—The four individuals examined were taken from the brackish waters of the Chilka Lake. Three of these were found upon the Sting Ray, *Trygon sephen*, Forsk. " attached outside, close to the junction of the skin and teeth on both upper and lower jaws."

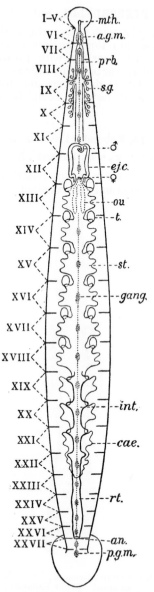

Fig. 18.—*Piscicola cæca*, Kaburaki, 1921. (After Kaburaki; the lettering slighty modified.) Diagram showing external and internal features. Somites numbered in Roman figures. *a.g.m.* Anterior ganglionic mass. *p.g.m.* Posterior ganglionic mass. *gang.* Ventral ganglion. *mth.* Mouth-opening. *prb.* Proboscis. *s.g.* Salivary glands. *st.* Stomach. *cæ.* Cæcum. *int.* Intestine. *an.* Anus. *ej.c.* Ejaculatory canal. *ov.* Ovary. *t.* Testis.

Genus **PTEROBDELLA**, Kaburaki, 1921.

Leeches inhabiting brackish water, ectoparasitic on fish. Body smooth and divided into three distinct regions, of which the anterior two are each provided with paired, lateral, fin-like processes. Without eyes or pulsating vesicles. Crop (or stomach) with five pairs of lateral diverticula and without a posterior cæcum. Five pairs of testes. Complete somite formed of fourteen rings (?).

7. Pterobdella amara, Kaburaki, 1921. (Figs. 19 & 20.)

Pterobdella amara, n. g., n. sp. Kaburaki, 1921, p. 668.

The brief notice here given of this remarkable species, which I have not had the opportunity of examining, is summarized from

Fig. 19.—*Pterobdella amara*, Kaburaki, 1921. Dorsal aspect, greatly enlarged. (After Kaburaki.)

Kaburaki's original description, and the diagnosis of the new genus which is added here, is based on information obtained from the same source.

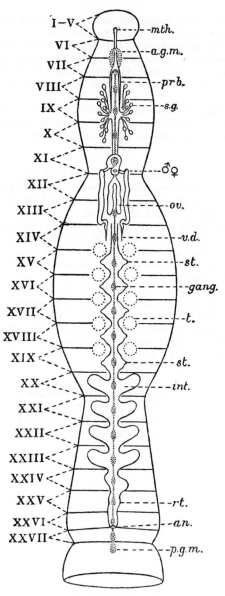

Fig. 20.—*Pterobdella amara*, Kaburaki, 1921. Diagram showing somites, ventral nerve-cord and the digestive and alimentary systems. (After Kaburaki.) Somites numbered in Roman figures. *mth*. Mouth. *a.g m*. Anterior ganglionic mass. *p.g.m*. Posterior ganglionic mass. *gang*. Ventral ganglion. *prb*. Proboscis. *s.g*. Salivary glands. *ov*. Ovary. *v.d*. Vas deferens. *t*. Testis. *st*. Stomach. *int*. Intestine. *rt*. Rectum. *an*. Anus.

Description.—Body depressed anteriorly, nearly circular posteriorly, and divided into three well-marked regions. The two anterior regions are each expanded laterally into paired, flattened, fin-like processes, a combination of features seen in no other known leech (see fig. 19). Ground-colour of the body white, occasionally with numerous minute pink spots on the dorsal surface.

Anterior sucker small, excentrically attached and somewhat campanulate. Posterior sucker a centrally attached, thick, circular disc about equal in width to the posterior part of the body.

No eyes.

Rings, in the individuals examined (which were preserved in alcohol), merged into irregular groups and not sufficiently distinct to render a correct count possible. Complete somite, as far as could be judged, formed of fourteen rings.

The reproductive and alimentary systems, ventral nerve-cord and distribution of somites are shown in fig. 20. The reproductive organs are of simple structure and open by a common pore situated a little in front of the division between somites XI and XII. Five pairs of testes.

The mouth opens in the centre of the anterior sucker, the œsophagus is unusually long and the crop or stomach is provided with but five pairs of lateral diverticula. The posterior extension of the crop, characteristic of the Ichthyobdellidæ, which takes the form of a cæcum or of paired cæca, is absent. The anus opens between somites XXVI and XXVII.

There are no pulsating vesicles.

Dimensions.—Length, 10 to 12 mm.; width, 2 to 3 mm.

Hosts and *Habitat.*—The examples of *Pterobdella amara* described were taken in the brackish waters of the Chilka Lake, from the Sting Rays, *Trygon sephen*, Forsk., and *Trygon uarnack*, Forsk. Usually the leeches were found firmly adhering to the gums of their hosts.

Family GLOSSIPHONIDÆ.

Clepsinidæ, Apáthy, 1888 *b*, p. 784.

Freshwater Rhynchobdellæ with ovate, flattened, never cylindrical body. Anterior sucker ventral and fused with the body. Posterior sucker cupuliform, distinct from the body, with a more or less ventral aspect. Crop (or stomach) aud intestine with conspicuous paired lateral cæca; the intestine always with four pairs. The eggs, enclosed in membranous sacs, are fixed and the young attach themselves to the ventral surface of the parent.

Genus GLOSSIPHONIA, Johnson, 1816.

Glossiphonia, Johnson, 1816, p. 25.
Glossopora, Johnson, 1817, p. 21.
Erpobdella, Blainville, in Lamarck, 1818.
Clepsine, Savigny, 1822.
Glossobdella, Blainville, 1828, p. 564.
Clepsina, Filippi, 1837.
Glossosiphonia, R. Blanchard, 1894, p. 24.

Glossiphonidæ generally of small size, with three or rarely with two pairs of eyes. Complete somite formed of three rings. Crop (or stomach) with six, or rarely with seven pairs of sublobate, lateral cæca, the last and longest pair reflected posteriorly. Mouth-opening within the anterior sucker.

8. Glossiphonia complanata, Linnæus, 1758. (Figs. 21 & 22.)

Hirudo complanata, Linnæus, 1758, p. 650 ; 1767, p. 1079.
Glossiphonia tuberculata, Johnson, 1816, p. 25.
Clepsine complanata, Savigny, 1822, p. 120.
Glossiphonia sexoculata, Moquin-Tandon, 1846, p. 353.
Glossiphonia cimiformis, Baird, 1869, p. 317.
Clepsine elegans, Verrill, 1872, p. 684.
Clepsine pallida, Verrill, 1872, p. 684.
Clepsine sexoculata, Apáthy, 1888 *a*, p. 154.
Glossosiphonia complanata, R. Blanchard, 1894, p. 27.
Glossiphonia elegans, Castle, 1900, p. 46.

(For complete synonymy and literature, see Harding, 1910, p. 158.)

Description.—The body of this well-known leech is ovate-elliptical, translucent aud generally of a dull green or brownish colour. Typically there are six longitudinal rows of yellowish spots which lie on the middle or sensory annulus of each somite, and correspond to the inner paramedian, intermediate and outer paramarginal papillæ and sense-organs.

The coloration and markings are very variable, but the species can usually be recognized by two conspicuous longitudinal dark brown interrupted lines upon the dorsal surface which appear in an inner paramedian position, the interruptions being due to the paramedian spots on each sensory ring. A pair of similar, though less conspicuous, dark lines occur ventrally, but rarely traverse the full length of the body. Apáthy (1888 *b*, p. 791) gives specific rank to a brownish form with six narrow longitudinal dark brown stripes (*G. concolor*) which, as Blanchard considered, is not more than a variety of *G. complanata*. Further investigation may lead to the subdivision of this widely-distributed species.

The six eyes lie in two close sub-parallel rows. The first and smallest pair, which sometimes may be absent, usually lie in the anterior part of somite III and are directed obliquely

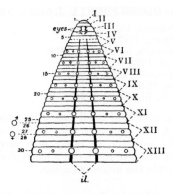

Fig. 21.—*Glossiphonia complanata*, Linn., 1758. Diagrammatic representation of anterior part of dorsal surface. Somites numbered in Roman and rings in ordinary figures. *i.l.* Interrupted dark lines.

forward. The second and largest pair, which are also pointed forward and to the side, lie in the posterior part of somite III. The last pair are directed obliquely backward and lie in the uniannulate somite IV.

Rings 68, of which two are usually preocular. Somites I–IV and XXVI–XXVII uniannulate, V–XXIV complete with three rings and XXV biannulate. Somite III shows signs of sub-division, and Moore (1924, p. 349) notes that this is more or less well marked in Indian examples. The first and second rings of somite V are usually imperfectly separated ; the first ring forms the posterior boundary of the anterior sucker.

The male genital orifice lies between somites XI and XII, and the female orifice lies between the second and third rings of the latter somite.

The anus opens in the anterior part of somite XXVII. There are ten pairs of testes, and the crop or stomach has six pairs

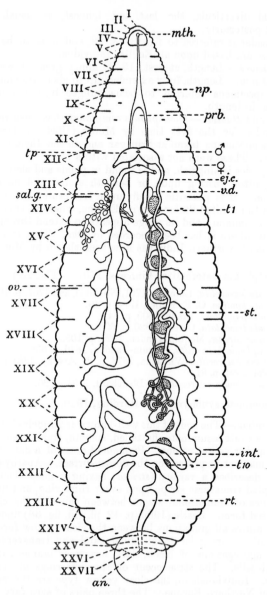

Fig. 22.—*Glossiphonia complanata*, Linn., 1758. Diagram showing alimentary tract, reproductive system and annulation. *mth.* Mouth-opening. *prb.* Proboscis. *sal.g.* Salivary glands. *st.* Stomach. *int.* Intestine. *rt.* Rectum. *an.* Anus. *np.* Nephridiopore. *ej.c.* Ejaculatory canal. *t.p.* Terminal portion of ejaculatory canal. *t 1, t 10.* First and sixth pair of testes. *ov.* Ovary. (After Harding; modified.)

of lateral diverticula, the last and longest, as usual, being reflected posteriorly.

The reader is referred to figs. 21 & 22, and it is to be noted that these are based upon an English individual.

Dimensions.—Length, at rest, 15–20 mm.; greatest width, at rest, 5–9 mm. Length, fully extended, up to about 35 mm. The Indian specimens referred to below ranged from 6·5 mm. to 21·2 mm. in length.

Hosts and *Habitat.*—*Glossiphonia complanata* has been recorded from India for the first time by Moore (1924, p. 348), who describes specimens taken at Srinagar and in the Jhelum Valley, Kashmir. These were not associated with a host.

The species is sluggish, and is found in ponds and slow-moving streams, often resting upon or beneath stones and on aquatic vegetation. It is parasitic chiefly upon freshwater snails. It is found in the United States and in Europe, where it is often exceedingly common; and its range appears to extend through Asia to Japan, where it has been recorded by Oka. That it wanders into parts of Northern India has now been shown by Moore.

9. Glossiphonia heteroclita, Linnæus, 1761. (Fig. 23.)

Hirudo heteorclita, Linnæus, 1761, No. 2085, and 1767, p. 1080.
Hirudo hyalina, O. F. Müller, 1774, p. 49.
Hirudo pappillosa, Braun, 1805, p. 64.
Hirudo trioculata, Carena, 1820, p. 303.
Clepsine carenæ, Moquin-Tandon, 1826, p. 105.
Glossiphonia heteroclita, Moquin-Tandon, 1846, p. 358.
Clepsine heteroclita, Whitman, 1878, p. 2; Apáthy, 1888 a, p. 154;
    Oka, 1894, p. 81.
Glossosiphonia heteroclita, R. Blanchard, 1894, p. 26.

For complete synonymy, see Harding, 1910, p. 155.

Description.—The body of this well-known and widely-distributed species is ovate acuminate, flattened, smooth, transparent, and of a clear amber-yellow colour. Pigmented areas of a darker colour may or may not be present on the dorsal surface. Apáthy (1888 b, p. 790) describes a variety (*striata*) having a deep black, often interrupted transverse stripe upon every third ring, and notes the occurrence of transitional stages between this and the clear unpigmented form. Castle (1900, p. 42, pl. viii, fig. 38) finds in the United States all gradations between the clear yellow form and a form with an irregular longitudinal band and transverve striæ, due to aggregations of black, dark brown or orange superficial pigment-cells. The striæ occur on the first rings of successive somites. Individuals of the clear yellow type are the most frequent in Northern Europe. The three pairs of eyes vary to some extent in position. The first pair usually lie in ring 5, and the second and third pairs are situated respectively in rings 7 and 8.

The first and smallest pair of eyes are closely approximated; in typical cases the right and left components of the second and third

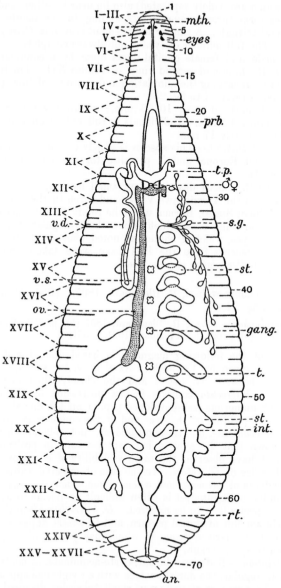

Fig. 23.—*Glossiphonia heteroclita*, Linn., 1761. Diagram showing annulation, alimentary tract and reproductive system. Somites numbered in Roman, and rings in ordinary figures. *mth.* Mouth-opening. *prb.* Proboscis. *s.g.* Salivary glands. *st.* Stomach. *int.* Intestine. *rt.* Rectum. *an.* Anus. *tp.* Terminal portion of ejaculatory canal. *v.s.* Vesicula seminalis or seminal reservoir. *v.d.* Vas deferens. *t.* Testis. *ov.* Ovary.

pairs (which are widely separated) also lie near together, and thus the eyes form three groups, corresponding to the points of an equilateral triangle. A similar arrangement is seen in *Glossiphonia weberi*. The full number of eyes is not always complete.

The male and female genital ducts open by a common pore between the first and second rings of somite XII. For further information regarding the annulation and anatomy of this species, the reader is referred to fig. 23.

Size.—Length at rest, 10–13 mm.; greatest width, at rest, approximately 4·5 mm. Individuals in full extension may attain a length of nearly 17 mm.

Hosts and *Habitat.*—This species is widely distributed in North America and Europe, and is parasitic for the most part upon Gasteropods.

Oka (1922, p. 522) describes four small leeches, of which one was taken in the Yawng-hwe Valley and three in the Inlé Lake, Southern Shan States, Burma, which he refers to this species, his identification being based chiefly upon the triangular disposition of the six eyes, which it shares, as noted above, with *G. weberi*. The three leeches from the Inlé Lake were taken from the Gasteropod, *Pachylabra maura*, Reeve.

10. Glossiphonia weberi, R. Blanchard, 1897. (Plate II, fig. 10; and fig. 24.)

Glossosiphonia weberi, R. Blanchard, 1897 (*b*), p. 332; Kaburaki, 1921 (*b*), p. 695, fig. 1; Moore, 1924, p. 351.

Much of the following diagnosis is based upon the detailed description given by Kaburaki. As that writer states, the real difference between *G. weberi* and the closely-allied and well-known *G. heteroclita* consists in the possession by the former species of numerous well-developed papillæ on the dorsal surface. The normal position of the eyes and of the common genital pore differs in each case, but these criteria are not reliable, since it is well known that the eyes in both species vary in position, and recently I have observed the same variability of situation in the genital orifice of *G. weberi*. In the case of material subjected to the accidents of preservation, when pigment may be washed out and papillæ obliterated, it is often difficult to decide to which of the two species it is to be referred. *G. heteroclita* often develops a certain amount of dorsal pattern, and *G. weberi* appears to be a tropical form derived from it which has just, and only just, attained to specific rank.

Description.—Body translucent, ovate-acuminate, in contraction nearly triangular, the dorsal surface with a roughened appearance, due to the presence of numerous small tubercles disposed transversely upon every ring.

In addition to these tubercles the dorsal surface bears a series of prominent metameric papillæ (see fig. 24), which form seven longitudinal rows extending from somite V to the posterior extremity. The papillæ composing six of these rows consist of a

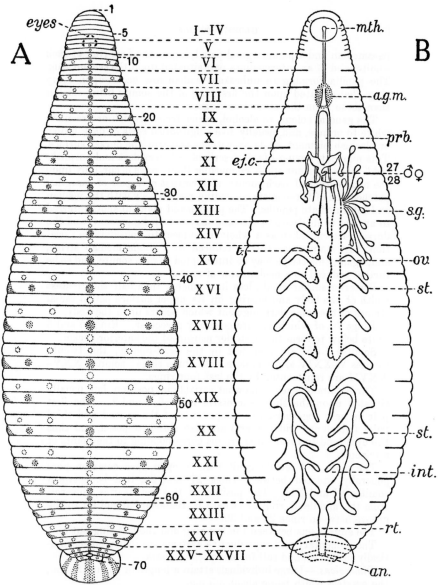

Fig. 24.—*Glossiphonia weberi*, R. Blanchard, 1897. Diagrams representing (A) dorsal aspect, showing external features, and (B) ventral aspect, showing reproductive organs, digestive tract, etc. (After Kaburaki; slightly modified.) The dorsal diagram (A) shows the seven longitudinal rows of papillæ in dotted outline, the dark pigment being indicated by dotted shading. Somites numbered in Roman, and rings in ordinary figures. *mth.* Mouth. *a.g.m.* Anterior ganglionic mass. *prb.* Proboscis. *s.g.* Salivary glands. *st.* Stomach. *int.* Intestine. *rt.* Rectum. *an.* Anus. *ej.c.* Ejaculatory canal. *t.* Testis. *ov.* Ovary. Certain parts are shown on one side only, for the sake of clearness.

paramedian and a paramarginal pair lying on the first ring of each somite, and of an intermediate pair situated upon the second ring. The papillæ forming an unbroken median row occur upon every ring, that situated upon the first ring of the somite being smaller than those upon the other two rings.

The general colour, in alcohol, varies from greyish or greenish-white to orange, and, usually, five longitudinal rows of dark brown or blackish pigment spots or patches traverse the dorsal surface. This pigment, during life, is described as " very dark and purplish-red." The pigment spots composing these rows occur only upon the middle ring of each somite, and consist of a pair of marginal patches and of three spots coinciding respectively with the median and intermediate papillæ. The papillæ and pigment spots referred to are apt to fluctuate in size and may not all be present, and the dorsal pattern, taken as a whole, is subject to considerable variation. In many cases this is reduced to a single median stripe, and, again, this stripe may be interrupted in each somite, and so be further reduced to a median row of spots.

Posterior sucker small, less in diameter than half the greatest width of the body and bearing on its upper surface paired radial stripes of the same dark pigment which occurs upon the body.

Rings 70, of which five are generally preocular. Rings 5 and 6 unite below to form the posterior margin of the anterior sucker.

Somites I–IV are absorbed by the head-region; the twenty somites V–XXIV are complete with three rings; somites XXV–XXVII are represented by the last four rings.

Three pairs of eyes, somewhat variable in position, but usually situated upon the three successive rings 6, 7 and 8. The first pair lie near together; the components of the second and third pairs (which are wider apart) are also closely apposed, and thus the eyes tend to form three groups corresponding to the points of a triangle.

The male and female genital ducts open by a common pore generally situated between rings 27 and 28, that is, between somites XI and XII. The position of this pore varies, however, from the middle of ring 27 (where R. Blanchard found it) to the groove between rings 28 and 29.

The anus opens between the last ring and the last ring but one.

The alimentary and reproductive systems are indicated schematically in fig. 24, and call for no special remark.

Dimensions.—Large individuals attain a length of about 12 mm., the greatest width being about 5·5 mm.

Hosts and *Habitat.*—*G. weberi* is widely distributed. R. Blanchard founded the species upon material received from Lake Manindjau, Sumatra, and it has since been recorded from (*a*) and (*b*) Bhim Tal, 4450 ft., Gurud Tal (near Sat Tal), 4550 ft., and (*c*) Naini Tal, 6300 ft., Kumaon, W. Himalayas; (*d*) a stream at Harwan, Kashmir; (*e*) Janikpur, Nepal; (*f*) Selai Kusi, Mangaldi, Assam; (*g*) N. end of Logtak Lake, Manipur; (*h*) Canal, Thantaung, W. side of Inlé Lake, Nyaungywe State, Burma;

(*i*) Lahore; (*j*) in and near Calcutta; (*k*) Diamond Harbour, R. Hoogli; (*l*) and (*m*) Cuttack, and Puri, Orissa; (*n*) S. end of Lake Chilka, N.E. Madras; (*o*) and (*p*) Chaibasa and Chakardharpur, Singbhum Dist., Chota Nagpur; (*q*) Burhanpur, Cent. Provinces; (*r*) Sangur, Cent. Provinces; (*s*) Itarsi, Hoshangabad Dist., Cent. Provinces; (*t*) Old bed of R. Narbada, N. of Babai, Hoshangabad Dist., Cent. Provinces; and (*u*) Whitefield, near Bangalore. *G. weberi* preys upon Gasteropods, the actual hosts hitherto recorded being species of *Ampullaria*, *Paludina* and *Limnæa*. It also attacks aquatic beetles, Dr. S. Kemp having found it upon the bodies of Dytiscidæ and upon a species of *Hydrophilus* at the end of a lake on Bhim Tal, Kumaon, W. Himalayas.

11. Glossiphonia reticulata, Kaburaki, 1921. (Fig. 25.)

The following brief diagnosis of this species is based upon the original description, given by Kaburaki, of a single individual in the possession of the Indian Museum.

Description.—Body slender, translucent, attenuated anteriorly, with the head-region somewhat dilated. Dorsal surface with a roughened appearance, due to the presence of numerous papillæ, and covered, like the ventral surface, with pigment, more or less reticulately distributed, which has faded in the preservative fluid to a uniform olive-grey hue.

Anterior sucker very small, less than half the width of the posterior sucker. Posterior sucker circular, nearly centrally attached and rather less than half the greatest width of the body.

Rings 72, of which three are preocular.

Somites I–IV are represented by the first six rings; the twenty somites V–XXIV are complete with three rings; somites XXV–XXVII are represented by rings 67–72.

Eyes two pairs; the first pair lie in ring 4, and the second and larger pair (in which the eyes are more widely separated) in ring 5.

Male genital orifice situated between rings 26 and 27 in somite XI; female orifice situated two rings behind the male, between rings 28 and 29, in somite XII.

The mouth opens within the anterior sucker, a little in front of the centre, and the anus lies between rings 70 and 71, being separated from the posterior sucker by two rings. The crop (or stomach) has 7 pairs of branching diverticula.

Dimensions.—Approximate total length, 11 mm.; approximate greatest width, 2 mm.

Host and *Habitat.*—The single specimen examined was found attached to the mantle of a species of *Anodonta* at Jullundhur.

12. Glossiphonia annandalei, Oka, 1922. (Fig. 26.)

The writer has not had the opportunity of examining examples of this remarkable species, and is indebted here to Oka's original description, which has been collated with the subsequent observations of Moore, 1924, p. 350.

F

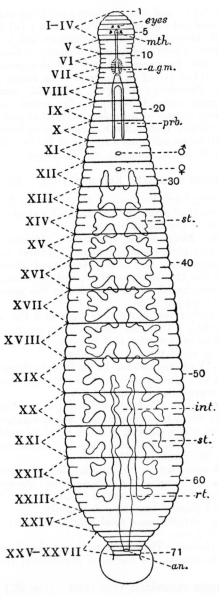

Fig. 25.—*Glossiphonia reticulata*, Kaburaki, 1921. (After Kaburaki; slightly modified.) Diagram showing external features and alimentary tract. Somites numbered in Roman, and rings in ordinary figures. *a.g.m* Anterior ganglionic mass. *mth*. Mouth-opening. *prb*. Proboscis. *st*. Stomach. *int*. Intestine. *rt*. Rectum. *an*. Anus.

Description.—Body elliptic-lanceolate, little flattened, having a nearly smooth surface devoid of conspicuous papillæ, and the head-region very slightly dilated. Oka's specimens, in alcohol, are described by him as being of a uniform pale grey colour. Moore's examples, on the other hand, were accompanied by the collector's notes on the coloration during life, which is described as "pale flesh-colour with minute dark dots on the dorsal surface, tending to run into hair-like lines."

Anterior sucker bounded posteriorly by the fifth ring and perforated by the small mouth-opening a little in front of the centre of its cup. Posterior sucker circular, and less in diameter than the greatest width of the body.

Rings 68, of which three are preocular. Somites I–III and XXVII uniannulate; IV biannulate, formed of an anterior broad

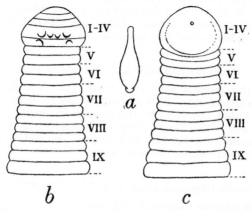

Fig. 26.—*Glossiphonia annandalei,* Oka, 1922. (After Oka.) *a.* Outline of entire leech, × 3. *b.* Somites I–X, dorsal view showing eyes, × 30. *c.* Somites I–IX, ventral view showing mouth-opening, × 30.

and a posterior narrow ring; V and VI also biannulate, but formed of rings of almost equal width. The eighteen somites VII–XXIV are triannulate, with rings of equal size; XXV and XXVI biannulate, with the anterior ring about twice as broad as the posterior.

The three pairs of eyes occupy a position unique among the Glossiphonidæ, and so provide a ready means of recognizing this species. On the posterior part of ring 4 lie two pairs of eyes, consisting of a small median pair situated between the components of a much larger pair. Another pair of large eyes lie on ring 5, immediately below the large eyes on ring 4. The openings of the pigment cups of all the eyes on ring 4 are directed forward, those of the eyes on ring 5 being directed backward. The pigment-cups of the large pairs of eyes lying respectively on rings 4 and 5 touch at their bases. This arrangement is shown in fig. 26.

The genital orifices are separated by two rings. The male pore opens in the furrow between rings 24 and 25, that is, between somites XI and XII, and the female pore opens between rings 26 and 27. There are six pairs of testes.

The seventeen pairs of nephridiopores lie between the first and second rings of the somites involved.

Crop with six pairs of diverticula, of which the first five pairs are simple and unbranched. The anus opens behind the last ring.

Dimensions.—The largest member of this species examined by Oka measured 6 mm. in length, its greatest width being 2·8 mm. Moore describes three examples of equal size which were 6·5 mm. long with a maximum width of 1·2 mm.

Hosts and *Habitat.*—Oka's specimens were found in the central region of the Inlé Lake, Southern Shan States, Burma, on a snail of the Family Viviparidæ, *Taia intha*, Annandale. The examples examined by Moore were Indian, having been taken in a freshwater pond on Samal Island, Chilka Lake, Madras Presidency. In this instance no host was recorded.

Genus **HELOBDELLA**, R. Blanchard, 1896.

Small Glossiphonidæ with one pair of eyes. Complete somite formed of three rings. Body generally without papillæ. Mouth-opening within the anterior sucker. Head-region continuous with the rest of the body. Crop (or stomach) with six pairs of simple lateral cæca, the last and longest reflected posteriorly. Sometimes with a dorsal chitinous scute.

13. **Helobdella stagnalis**, Linnæus, 1758. (Fig. 27.)

Hirudo stagnalis, Linnæus, 1758, p. 649.
Clepsine bioculata, Moquin-Tandon, 1826, p. 102.
Glossosiphonia stagnalis, R. Blanchard, 1894, p. 25.
Helobdella stagnalis, R. Blanchard, 1896, p. 4.

[For complete synonymy, see Harding, 1910, p. 162.]

Owing to the courtesy of Dr. H. A. Bayliss, I was enabled, in 1922, to examine an individual of this species which had been sent to the British Museum of Natural History from the Himalayas.

Description.—This widely distributed species, here recorded from India for the first time, has a slender translucent body of a uniform greenish, yellowish or brownish hue finely speckled with black. A dorsal circular chitinous plate or scute, situated between the twelfth and thirteenth rings, forms the most striking external feature. A similar scute is present in the South American species *Helobdella scutifera* (R. Blanchard, 1900). Posterior sucker small, its width being not greater than about half the greatest width of the body.

Rings 68. The somites are not readily distinguished by external bservation. Somites I–V are represented by the first five rings;

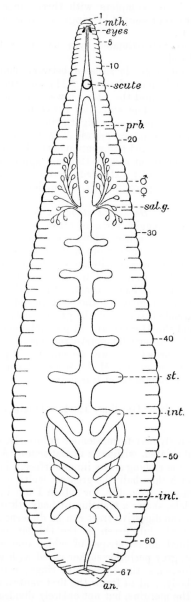

Fig. 27.—*Helobdella stagnalis*, Linn., 1758. (After Harding.) Diagram showing alimentary tract. *mth.* Mouth-opening. *prb.* Proboscis. *sal.g.* Salivary glands. *cr.* Crop. *st.* Stomach. *int.* Intestine. *an.* Anus.

somites VI–XXIV are complete with three rings, and somites XXV–XXVII are represented by rings 63–67 (see Castle, 1900, p. 22, pl. ii, fig. 4).

The two closely approximated eyes lie in the third ring or between rings 2 and 3.

The male genital orifice is situated between rings 24 and 25 (the first and second rings of somite XII), and the female orifice is situated one ring behind the male, between the second and third rings of somite XII. The anus lies between the sixty-seventh and the last and incomplete sixty-eighth ring.

The first five pairs of lateral gastric cæca, when undistended by food, may become retracted and consequently difficult to detect.

Dimensions.—Length, at rest, 8–12 mm.; approximate width, at rest, 4 mm.; length, fully extended, as much as 26 mm. The Indian example was approximately 6 mm. long. and 2·5 mm. in width.

Hosts and *Habitat.*—*Helobdella stagnalis* is met with in lakes, ponds, ditches and sluggish streams, often resting upon aquatic plants, and although parasitic chiefly upon Gasteropods, it preys also upon a considerable variety of small freshwater invertebrates, and has been noticed occasionally on the bodies of frogs, newts and injured fish. It is found in Canada and in the United States from the Atlantic to the Pacific coast. In South America it has been recorded from Paraguay and the western slopes of the Andes, and its range extends throughout the greater part of Europe into Western Asia and the Himalayas. It probably occurs throughout the northern and southern temperate regions of the globe. The Indian example recorded here was accompanied by a note stating that it had been taken by Mr. M. E. Moseley from a stream at an altitude of 11,000–12,000 ft. in Kashmir.

14. Helobdella nociva, Harding, 1924.   (Plate II, fig. 9; and fig. 28.)

Description.—Body claviform, slender anteriorly, with the head-region somewhat dilated, without a dorsal scute. Anterior sucker with fine transverse ribbing on its inner surface. Posterior sucker small, less in width than half the greatest width of the body. A water-colour drawing from life, for which I am indebted to the late Dr. Annandale, shows a translucent body (through which the dark contents of the digestive tract can be discerned), of a pinkish hue where thinnest, dull green in the thicker parts, with five brown, longitudinal dorsal stripes of which one is median, and having the bluish-grey posterior sucker rayed with white.

Rings 70. There are three preocular rings; ring 9 is the first to encircle the body, and so forms the first ventral annulus; ring 68 is double at the margins, but not entirely divided.

The single pair of eyes occupy ring 4.

Two dorsal pairs of papillæ (all that could be made out in the individuals examined) situated one pair on either side of the

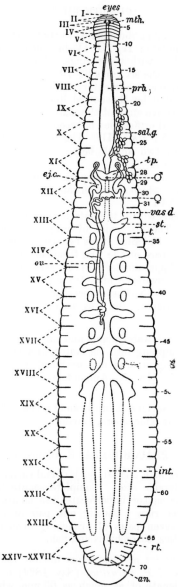

Fig. 28.—*Helobdella nociva*, Harding, 1924. (After Harding.) Diagram
showing external features and the reproductive and alimentary systems.
Somites numbered in Roman, and rings in ordinary figures. *mth*. Mouth-
opening. *prb*. Proboscis. *sal.g*. Salivary glands. *ej.c*. Ejaculatory canal.
t.p. Terminal portion of ejaculatory canal. *vas.d*. Vas deferens. *t*. Testis.
ov. Ovary (shown on side only for the sake of clearness). *st*. Stomach. *rt*.
Rectum. *int*. Intestine. *an*. Anus. The dotted portion of the alimentary
tract could not be seen owing to the imperfect state of the material.

median line, distinguish the middle or primary ring of each somite.
Somite I uniannulate ; II, III IV biannulate ; the 19 somites
V–XXIII complete with three rings.   Somites XXIV–XXVII
are represented by the six rings 65–70.

The reproductive and alimentary systems are shown in fig. 28.
(The parts indicated by the dotted lines were not clearly seen,
owing to the defective state of the material.)

The male genital orifice lies between rings 28 and 29, that is,
between somites XI and XII.   The female orifice is separated
from the male by two annuli, being situated between rings 30
and 31 in somite XII.

The genital organs call for no special comment, with the exception of the ovaries, which are much coiled and very long, extending
as far as somite XVI.

The mouth opens within the anterior sucker a little in advance
of the centre.   The six pairs of crop (or stomach) cæca are
somewhat lobed.   The anus opens between rings 69 and 70, being
separated by the space of one ring from the posterior sucker.

Dimensions.—Approximate size of largest individual: total
length, 7·5 mm. ; greatest width, 1·5 mm.

Host and *Habitat.*—No host is recorded.   The material examined
came from two sources, and was accompanied by the following
notes :—

(*a*) " On stems of water plants and on under surface of *Canna*
        leaves dipping into the water, in a small pond of fresh
        water dug in sand and overshadowed by trees at Puri,
        Orissa."   (Dr. N. Annandale and Dr. F. H. Graveley
        coll.)

(*b*) " On under surface of bricks and broken earthenware pots in
        a tank at Kidderpore, Calcutta."   (R. Hodgart coll.)

Genus PLACOBDELLA, R. Blanchard, 1893.

Body flattened, with a crustaceous dorsal surface and sometimes
attaining considerable size.   Complete somite formed of three
rings.   Anterior sucker imperforate, the mouth being situated
upon its anterior rim.   Usually one pair of eyes.   Crop (or
stomach) with seven pairs of branching diverticula.   Parasitic
chiefly upon turtles, batrachians and fish.

The largest species of this genus inhabit the United States,
where examples of *P. parasitica* (Say, 1824) may be met with
which attain a length of as much as 60 mm. when at rest, and
more than 80 mm. when fully extended.   The eyes of several of
these American species, when subjected to refined methods of
examination, have been found to be compound.   The crustaceous
or roughened dorsal surface seen in members of this genus is due
to the presence of numerous small cutaneous papillæ closely set
upon every ring, and the terminal mouth-opening, although not

peculiar to *Placobdella*, is somewhat uncommon and an easily recognized and valuable character.

The Indian species of *Placobdella* at present definitely known are of comparatively small size, and probably attack a wider range of hosts than are here recorded.

15. Placobdella ceylanica, Harding, 1909.

Glossiphonia ceylanica, Harding, 1909, p. 233.
Glossosiphonia ceylanica, Kaburaki, 1921 *a*, p. 671, fig. 5.
Placobdella ceylanica, Moore, 1924, p. 357, pl. xix, fig. 7 and pl. xxi, fig. 25.

In 1909 I received a single, imperfectly preserved example of this little leech, and published a brief, preliminary description of it which has proved to be not free from error. In 1921, Kaburaki, working upon further material from India, was able to make a more complete examination of the species, and his results, in 1924, were revised by Moore, who had received additional Indian examples. Thus, as I am glad to acknowledge, the credit of establishing this species is really due to these two authorities, upon whose work the following description is based. I here follow Moore in referring this species, provisionally at least, to the genus *Placobdella*, on account of the terminal oral opening and the seven pairs of gastric cæca ; and notwithstanding the three pairs of eyes.

Description.—Body lanceolate, smooth, flattened, with the head-region slightly dilated. Posterior sucker small, its diameter being equal to about half the greatest width of the body.

Colour in alcohol pale buff, grey or brown, somewhat lighter below ; dorsal surface with three longitudinal dark brown lines or rows of spots.

Mouth-opening very small, at the extreme anterior margin of the anterior sucker.

Rings 71, of which two are preocular.

Somites I uniannulate ; II biannulate (the groove between the two rings shallow) ; III perfectly biannulate, the twenty-one somites IV–XXIV being complete with three rings. Somite XXIV has the third ring reduced, XXV is biannulate and XXVI and XXVII are uniannulate.

Six eyes disposed in two subparallel rows. The first and second pairs of eyes lie respectively in rings 3 and 4 ; the third pair occur on the sixth ring (the sensory ring of somite IV). The separation of the second and third pairs of eyes by two rings is characteristic of the species.

Male genital orifice situated between somites XI and XII ; female orifice two rings behind the male, between the second and third rings of somite XII.

Crop (or stomach) with seven pairs of cæca, the last and longest reflected, as usual, posteriorly. The first six pairs and the lobes of the last pair bifurcate at their extremities.

The nephridiopores pierce the middle ring of the somite, but their total number could not be ascertained in the material examined. The anus lies between the penultimate and the last ring.

Dimensions.—At rest, length approximately 8–13 mm.; greatest width, approximately, 3 mm.

Hosts and *Habitat.*—*G. ceylanica* was first recorded from Ceylon, on the Mud-turtle, *Emyda granosa vittata.* Kaburaki notes its occurrence in India at Rawalpindi, and on Barkuda Island, Chilka Lake, Madras Presidency, in a pond where *Rana cyanophlyctis*, Schmid., was the host. Moore records it from a small freshwater pond on Samal Island, Chilka Lake, off *Emyda granosa intermedia*, and also from a tank in the Government Garden, Buldana, Central Provinces.

16. Placobdella emydæ, Harding, 1920. (Fig. 29.)

Placobdella emydæ, Harding, 1920, p. 514 ; Kaburaki, 1921, p. 701.

Description.—Body with the characters of the genus; in extension elliptic-lanceolate ; with the head-region somewhat dilated. The middle ring of the somite bears dorsally three pairs of metameric papillæ : a paramedian, an intermediate and a paramarginal pair, the intermediate pair being the largest.

The ground-colour varies from greyish-green to pale olive-brown, the gorged crop or stomach (in the individuals examined) appearing dull green through the translucent body. Dorsal surface with a white median stripe and profusely speckled with white and a darker green. Ventral surface smooth and olivaceous.

Anterior sucker with a shallow interior cup having a finely-ribbed surface somewhat resembling that of the human finger-tip. Posterior sucker circular, centrally attached, narrower than the widest part of the body, its upper surface bearing paired white rays.

Rings 71. The second and third rings are each subdivided at their margins ; the fifth is confluent with the free ventral posterior edge of the anterior sucker. Rings 6 and 7 are sometimes so slightly divided above as to give the appearance of a single ring ; the former disappears ventrally, leaving ring 7 to form the first ventral ring behind the anterior sucker.

Somites I–III uniannulate; IV, XXV, XXVI and XXVII biannulate ; the twenty somites V–XXIV complete with three rings.

The single pair of eyes usually lie in the third ring, but sometimes appear between rings 2 and 3.

Male genital opening situated between rings 26 and 27, that is, between somites XI and XII. Female orifice two rings behind the male, between rings 28 and 29, in somite XII.

The more important external and internal features are shown in fig. 29. The anus opens between rings 69 and 70, and thus is separated by two rings from the anterior sucker. There are six

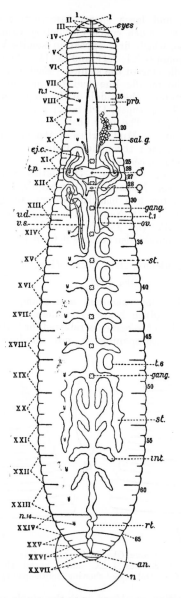

Fig. 29.—*Placobella emydæ*, Harding, 1920. (After Harding.) Diagram showing external and internal features. Somites numbered in Roman, and rings in ordinary figures. *prb.* Proboscis. *sal.g.* Salivary glands. *ej.c.* Ejaculatory canal. *t.p.* Terminal portion of ejaculatory canal. *v.s.* Vesicula seminalis or seminal reservoir. *v.d.* Vas deferens. *gang.* Ventral ganglion. *st.* Stomach. *rt.* Rectum. *an.* Anus. 1, *n.*14. First and fourteenth pair of nephridiopores (on ventral surface *n.*

pairs of testes, and large sperm reservoirs connect the vasa deferentia and the much coiled ejaculatory canals. Nephridia 14 pairs. These, which first appear in somite VIII, are absent in the clitellar somites XI, XII and XIII. The nephridiopores open in the middle ring of the somite. Each pore lies about mid-way between the middle line and the margin of the ventral surface, and perforates the middle of the ring in which it appears.

Dimensions.—Total length, 13·5 mm.; greatest width, 9 mm.

Hosts and Habitat.—The only hosts upon which this leech has so far been found are Mud-turtles of the genus *Emyda*. It is probably fairly common throughout India, and has been recorded from the following places :—

(*a*) Outskirts of Calcutta. (*b*) Gatiagurgh Dist., Hoogli, Bengal. (*c*) River Mahanaddi, Sambalpur, Orissa. (*d*) The Chilka Lake [in the less blackish waters, sp. g. 1·007–1·011]. (*e*) Near Purulia, Chota Nagpur Div. (*f*) Nagpur, C.P. (*g*) Hoshangabad, C.P. (*h*) Taloshi (c. 2000 ft.), Koyna Valley, Satara Dist. Kaburaki also records it from Burma.

17. Placobdella inleana, Oka, 1922. (Fig. 30.)

Glossiphonia inleana, Oka, 1922.

The writer has not examined any example of this species and the diagnosis given below is a summary of Oka's original description, to which the reader is referred for further information. Although the mouth-opening is not absolutely terminal, its

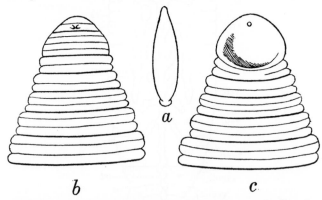

Fig. 30.—*Placobdella inleana*, Oka, 1922. (After Oka.) *a.* Outline of entire leech, × 3. *b.* Somites I–IX, dorsal aspect showing eyes, × 30. *c.* Somites I–X, ventral aspect showing mouth-opening, × 30.

subterminal position, combined with the single pair of eyes and the seven pairs of crop diverticula, seem to indicate that this species is more nearly related to the genus *Placobdella* than to *Glossiphonia*, and it is referred here, at least provisionally, to the former genus.

Description.—Body ovate-oblong, somewhat convex both above and below, with the lateral margins sharply serrate. The examples examined were preserved in alcohol and had faded to a uniform pale grey colour with, however, indications of roundish spots arranged in regular series on the dorsal surface.

Anterior sucker bounded posteriorly by ring 6, with the mouth-opening situated in a subterminal position, immediately below the eyes. Posterior sucker about half the width of widest part of the body.

One pair of large and distinct eyes lie upon the second ring, with their bases back to back and nearly touching, the openings of their pigment-cups being pointed in a lateral and slightly forward direction.

The number of rings in adult individuals is usually 67 ; the annulation, however, presents remarkable peculiarities.

In this species, as in certain others, the somite limits are readily recognizable externally, owing to the furrows separating the rings of contiguous somites being more conspicuous than the other interannular furrows.

The interannular furrows of the three-ringed somites are not all of the same depth, the groove separating the first and second rings being shallower than that separating the second and third. Again, in most cases, the three rings of a somite are of different widths, the middle ring being the widest, the last ring somewhat less wide and the first ring always the narrowest. Some of the biannulate somites also have rings of unequal width, the anterior ring, in these cases, being the largest. Finally, the number of rings increases to some extent with growth, owing to certain rings which are single in the young leech becoming divided in the adult. It is possible, in this species, " to observe the various stages through which the primitive uniannulate somite of the ancestral leech gradually became the typical triannulate somite of the Glossiphonidæ."

Somites I, II, XXVI and XXVII are always uniannulate; V, VI and XXIV are always biannulate, and IX–XXII are always complete with three rings. The remaining somites vary in annulation with the growth of the individual, viz. III, IV and XXV are uniannulate in the young leech and biannulate in the adult; similarly VII, VIII and XXIII are originally biannulate, but become triannulate as full size is attained.

Genital openings separated by two rings. In adult fully annulated individuals the male pore lies between rings 26 and 27, and the female pore between rings 28 and 29.

There are six pairs of testes, sixteen pairs of nephidia and seven pairs of gastric diverticula.

Dimensions.—Full-grown individuals measure about 9 mm. in length and nearly 3 mm. in width.

Host and *Habitat.*—About sixty examples were taken from a tortoise, *Cyclemys dhor shanensis*, Annandale, at Fort Stedman, Inlé Lake, Southern Shan States, Burma.

18. **Placobdella fulva,** Harding, 1924. (Plate II, fig. 7; and
fig. 31.)

I am indebted to the late Dr. Annandale for information
regarding the colour during life of this elegant little species.

Description.—Body flattened, in extension claviform and very
slender anteriorly. Upper surface of a bright reddish-yellow hue,
having in addition to the deep brown markings described below
a conspicuous longitudinal median cream-coloured stripe and a pair
of broken, marginal cream-coloured bands. Ventral surface white.
Dorsally each ring bears a large median papilla, and these papillæ
form a prominent row corresponding to the median cream-coloured
stripe. An intermediate and a marginal pair of papillæ present
on the dorsal surface of the middle ring of each somite are
covered by deep brown spots, which are connected longitudinally
by dark brown lines. Head-region undilated and continuous
with the body. Posterior sucker small, centrally attached, not
wider than half the greatest width of the body, with paired
cream-coloured rays.

Rings 67. Ring 5 forms the posterior boundary of the anterior
sucker and is the first to appear on the ventral surface. Ring 63
is doubled at the margins but not entirely divided.

Somites V–XXII complete with three annuli.

The single pair of eyes lie within ring 2, there being one
preocular ring.

The male genital pore is situated between rings 26 and 27, that
is, between the first and second annuli of somite XI; the female
pores open between rings 28 and 29, that is, between somites XI
and XII.

The anus lies between rings 66 and 67, being separated by one
ring from the posterior sucker.

(The accidental destruction of the two individuals upon which
this species is founded prevented the complete investigation of
their internal features.)

Dimensions.—Approximate total length, 13 mm.; approximate
greatest width, 2 mm.

Hosts and *Habitat.*—There is no positive record of a host. The
leeches examined were stated to have been found on the lower
surface of a dead *Unio* shell at the edge of a stream, at Purulia,
Manbhum Dist., Chota Nagpur Div., Bengal ("N. Annandale
and F. H. Graveley coll.").

19. **Placobdella undulata,** Harding, 1924. (Fig. 32.)

Description.—The typical elliptic lanceolate form of the body is
modified in adult individuals by a slight constriction centred at
the thirtieth ring, immediately behind the female orifice. Dorsal
surface with a roughened appearance, due to the presence on each
ring of numerous closely-set papillæ. Head-region somewhat
dilated and distinct from the body. Posterior sucker circular of

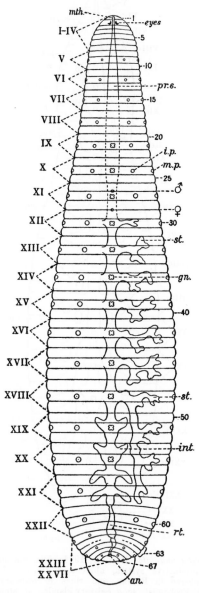

Fig. 31.—*Placobdella fulva*, Harding, 1924. Diagram showing external features, digestive tract and nerve-ganglia. Somites numbered in Roman, and rings in ordinary figures. *ip.* Intermediate papilla. *mp.* Marginal papilla. *mth.* Mouth. *prs.* Proboscis sheath. *st.* Stomach. *int.* Intestine. *rt.* Rectum. *an.* Anus. *gn.* A ganglion of the ventral chain. (After Harding.)

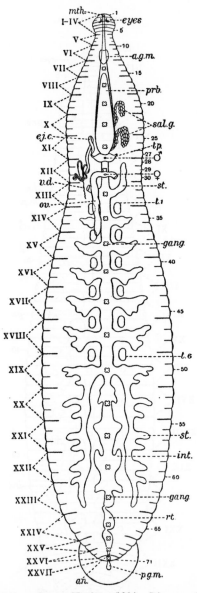

Fig. 32.—*Placobdella undulata*, Harding, 1924. Diagram showing external
features, reproductive system, alimentary tract and nerve-ganglia.
Somites numbered in Roman, and rings in ordinary figures. *mth.*
Mouth. *prb.* Proboscis. *sal.g.* Salivary glands. *st.* Stomach. *int.* Intestine. *rt.* Rectum. *an.* Anus. *ej.c.* Ejaculatory canal (shown only
on one side for the sake of clearness). *t.p.* Terminal portion of ejaculatory canal. *v.d.* Vas deferens. *t.1, t.6.* First and sixth pair of testes.
ov. One of the ovaries. *a.g.m.* Anterior ganglionic mass. *p.g.m.* Posterior
ganglionic mass. *gang.* A ganglion of the ventral chain. (After Harding.)

slightly oval, small, centrally attached, about equal in width to half the greatest width of the body. Colour faded in alcohol to a uniform buff hue.

Rings 71. Ring 2 is double at the margins but not entirely divided.

Somites I–IV, which could not be plotted with certainty in the material examined, appear to be represented by rings 1–6, which overlie the anterior sucker. Somites V–XXIV complete with three annuli; XXV and XXVI biannulate; XXVII uniannulate. Rings 7 and 8, the first and second rings respectively of somite V, unite below to form the first ventral ring; somite V, therefore, is only triannulate dorsally.

There is one preocular ring; the single pair of eyes lie in the second ring.

Male genital orifice situated between rings 27 and 28, that is, between somites XI and XII. The female orifice lies two rings behind the male, in somite XII, between rings 29 and 30. The anus opens between rings 70 and 71, thus being separated by one ring from the posterior sucker.

The reproductive and alimentary systems are shown in fig. 32, and conform in all important particulars to the general plan characteristic of this genus. The mouth opens in the usual terminal position; the salivary glands form paired compacted masses; the seven pairs of gastric cæca are moderately lobate; the paired pouches of the intestine are somewhat less developed than usual.

Dimensions.—Approximate size of the largest example: total length, 17·5 mm.; greatest width, 4 mm.; width at the deepest point of constriction (ring 30), 2 mm.; greatest width of head-region, ·75 mm.; width of posterior sucker, 2 mm.

Host and *Habitat.*—I am greatly indebted to Prof. Clifford Dobell, F.R.S., for the material upon which this species is founded. A note enclosed with the leeches examined states that they were said to have been taken from Koraliya fish (*Etroplus suratensis*) in the Colombo Lake, Ceylon.

Genus THEROMYZON, Philippi, 1867.

Glossiphonia, Johnson, 1816 (partim).
Clepsine, Savigny, 1822 (partim).
Hæmocharis, de Filippi, 1837 (partim); not *Hæmocharis,* Savigny, 1822.
Hemiclepsis, Vejdovsky, 1883 (partim).
Protoclepsine, Moore, 1898.
Protoclepsis, Livanow, 1902.

Glossiphonidæ of medium size, with four pairs of eyes. Complete somite formed of three rings. Somite III is rarely and somites IV–XXIV are always complete. The crop (or stomach), which has more than seven pairs of lateral diverticula, extends anteriorly into the preclitellar region.

These are generally slender, elongate, soft, delicate, somewhat flattened leeches, of a more or less greyish-green or brown colour, with yellow spots upon the upper surface of the body and also of the posterior sucker, where they are disposed in a marginal series. During life they are often possessed of great powers of extension and contraction, and of extraordinary restlessness and activity, creeping rapidly with a looping movement upon the slightest disturbance. Most of the known species have been found in Lake Baikal, in Siberia, but two species, at least, have a wider distribution, namely the one under discussion and *T. tessellata* (O. F. Müller, 1774).

I here follow Moore (1924, p. 346) in giving Philippi's name *Theromyzon* precedence of the more familiar name *Protoclepsis* applied by Livanow to this genus, which he was the first to place upon a satisfactory basis. Livanow divides the genus into two groups, distinguished by the following characters :—

(*a*) Genital orifices separated by two rings. In adults there is a shallow primitive form of vagina opening to the exterior by a single female pore. In undeveloped individuals there is no vagina and the oviducts open directly to the exterior, where they appear (in material which has been sectioned) as a pair of very small apertures situated one on either side of the median line.

(*b*) Genital apertures separated by more than two rings. The single female pore opens into a well-developed vagina.

The species referred to below belongs to the first of these groups, which contains but one other species, *T. garjaewi* (Livanow, 1902).

20. Theromyzon sexoculata, Moore, 1898.

Protoclepsine sexoculata, Moore (1898, p. 546).
Protoclepsis meyeri, Livanow (1902, p. 345).

Moore (1924, p. 346) refers two specimens of *Theromyzon* from Manipur somewhat doubtfully to this species, with which they agree "in most features of annulation, the position of the genital pores, etc."

Description.—According to Livanow (*loc. cit.*), who describes it under the name of *Protoclepsis meyeri*, this species has an elongate slender body, convex above and flattened below, and of an olive-green colour. The dorsal surface is traversed by six longitudinal rows of yellow spots. The spots composing four of these rows occur on the middle or sensory ring of each somite, and correspond to the outer paramedian and intermediate papillæ. (The inner paramedian papillæ are absent.) The spots forming the two remaining rows lie in an inner paramarginal position on the last ring of each somite.

The annulation is similar to that of the typical form, *Theromyzon* (*Protoclepsis*) *tessellata*, with which it also agrees in the position of the eyes.

Somite I uniannulate; II usually biannulate, the second ring. however, being occasionally missing. Somite III so rarely tri-annulate that it may be regarded as typically biannulate, its first and second rings being nearly always merged together. Somites IV-XXIV complete with three rings, and XXV-XXVII biannulate.

Rings (when somite III is biannulate) 74 in number.

Eyes, four pairs (forming two subparallel rows), situated respectively upon rings 2, 4, 7 and 10 (on the sensory rings of the somites II, III, IV and V, in which they respectively lie).

The male genital pore lies between somites XI and XII, and the female pore two rings behind it, between the second and third rings of XII. The anus opens between rings 73 and 74, in the middle of somite XXVII.

Dimensions.—9 mm. long, with a maximum width of 3·1 mm. (*Moore*); 8 mm. long and 3 mm. wide (*Livanow*).

Hosts and *Habitat.*—One of the two Indian examples of *T. sexo-culata* noted by Moore came from Loktak Lake, Manipur, Assam; the other was found in a small stream flowing out of this lake. Moore's original description of this species (1898) was based upon an example from Behring Island, Commander Islands, Siberia. Livanow's specimens came from Russia, and he notes the occurrence of this leech also in France and Sweden. The only host mentioned is Wild-duck. Probably, as in the case of *T. tessellata*, other waterfowl are also attacked.

Genus HEMICLEPSIS, Vejdovsky, 1883.

Glossiphonidæ of medium size, typically with two pairs of eyes. Complete somite formed of three rings. Head-region dilated and distinct from the rest of the body. The crop (or stomach), which has more than seven pairs of lateral diverticula, extends anteriorly into the preclitellar region. Mouth-opening within the anterior sucker.

21 *a*. **Hemiclepsis marginata**, subspecies **marginata**, O. F. Müller, 1774. (Plate II, figs. 1 & 2; and figs. 33 & 34.)

Hirudo marginata, O. F. Müller, 1774, p. 45.
Glossiphonia marginata, Moquin-Tandon, 1846, p. 375, pl. xiv, figs. 10-20.
Hemiclepsis marginata (in India), Kaburaki, 1921, pp. 694-695.

(For full synonymy and literature, see Harding, 1910, pp. 151-152.)

Description.—The flattened claviform and translucent body is usually richly pigmented, but the coloration is subject to con-siderable variation both in kind and in intensity. In typical examples the thin margins and extremities are colourless or hyaline and the ground-colour of the thicker parts is yellow, profusely sprinkled above with bright green. When, however,

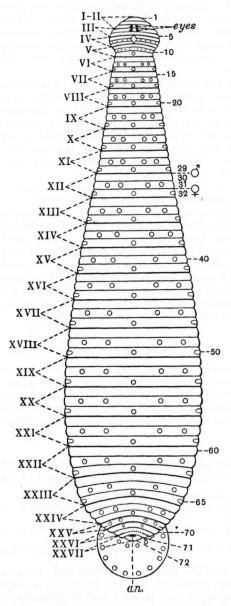

Fig. 33.—*Hemiclepsis marginata*, subsp. *marginata*, O. F. Müller, 1774. Diagram showing external features. Somites numbered in Roman, and rings in ordinary figures. Pigmented spots indicated in circular outline. *an.* Anus. (After Harding.)

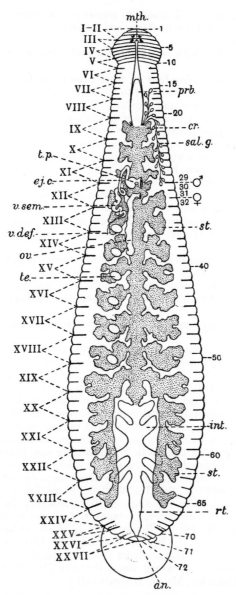

Fig. 34. *H. marginata*, subsp. *marginata*, O. F. Müller, 1774 Diagram
showing alimentary tract and reproductive organs. The latter, and also
the salivary glands, are shown on one side only for the sake of clearness
mth. Mouth. *prb.* Proboscis. *sal.g.* Salivary glands. *st.* Stomach.
(shaded). *int.* Intestine. *rt.* Rectum. *an.* Anus. *ej.c.* Ejaculatory canal
t.p. Its terminal portion. *v.sem.* Vesicula seminalis or seminal reservoir
v.def. Vas deferens. *te.* Testis. *ov.* Ovary. (After Harding.)

the crop is gorged with blood its scarlet hue, shining through the
semi-transparent body modifies, and sometimes entirely eclipses,
the green pigment.

Dorsal surface with seven longitudinal rows of lemon-yellow
spots.

The spots composing four of these rows lie on the middle ring
of each somite and correspond to the outer paramedian and intermediate sense-organs and papillæ. The spots forming a median
row fall upon the third ring, and the two remaining series, which
occupy a marginal position, occur upon the third, and sometimes
also upon the middle ring of each somite.

Posterior sucker with an outer, and often with an inner series
of lemon-yellow spots between which reddish-brown radial stripes
are often present.

Rings 72, of which two are preocular. The seventy-first is
partly subdivided and traces of subdivision appear in ring 69.
The twenty-one somites IV–XXIV are complete with three rings.

The two pairs of eyes are situated respectively upon the third
and fourth rings.

The male genital orifice opens between somites XI and XII,
that is, between rings 29 and 30, and the female orifice lies two
rings behind the male, between the second and third rings of
somite XII. The anus opens between the last and the penultimate ring.

The alimentary tract and the reproductive organs are indicated
in fig. 34.

Dimensions.—Length, at rest, 15–20 mm.; width, at rest,
3–7 mm.; length, fully extended, up to 30 mm.

Hosts and *Habitat.*—*Hemiclepsis marginata marginata* is chiefly
a fish parasite, but it also attacks certain molluscs, and has been
taken in India from a species of *Lamellidens* (by Dr. T. Southwell). It inhabits freshwater ponds, streams and lakes, where
it is often found upon water-plants and various other submerged
objects, lying in wait for its prey. Its range extends throughout
the greater part of Europe to Western Asia and India, where it
begins to be replaced by the subspecies *asiatica*. In India it has
been recorded from the following localities:—(*a*), (*b*), (*c*), (*d*)
Malwa Tal, 3600 ft., Sat Tal, 4500 ft., Bhim Tal, 4450 ft., and
Naukuchia Tal, 4200 ft., Kumaon, W. Himalayas (Dr. S. Kemp
coll.); (*e*), (*f*), (*g*) Janikpur, Chitauni, and Chukei Mukei, Nepal;
(*h*) Igatpuri, W. Ghats, Bombay P.; (*i*) Puri, Orissa (Dr. N.
Annandale coll.); (*j*) Bagra, Hoshangabad Dist. (Dr. F. H.
Graveley coll.); (*k*) Bhandardaha Beel, Murshidabad Dist. (Dr. T.
Southwell coll.); (*l*) in and about Calcutta (Dr. N. Annandale,
Dr. F. H. Graveley, Mr. R. Hodgart coll.).

21 *b.* **Hemiclepsis marginata**, O. F. Müller, subspecies **asiatica**, Moore, 1924. (Fig. 35.)

Moore (1924, p. 359) describes a subspecies of *Hemiclepsis marginata* which he names *asiatica* and states to be the most abundant and generally distributed of the Glossiphonidæ in Kashmir. He regards this as a form intermediate between the typical *H. marginata* of Europe, with its four well-developed eyes, and Oka's species, *H. casmiana*, from China and Japan, which has only two eyes, but resembles the subspecies *asiatica* in colour and dorsal pattern. He suggests, further, that *H. marginata* should be divided into three subspecies, viz. *marginata*, representing the typical form, *casmiana* the Far Eastern form, and *asiatica* the form described here. This suggestion, as far as the Indian region is concerned, has been adopted in these pages.

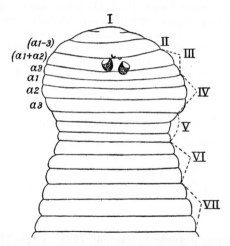

Fig. 35.—*Hemiclepsis marginata*, subsp. *asiatica*, Moore, 1894. (After Moore.) Dorsal view of head-region showing eyes and annulation. Somites numbered in Roman, and rings in ordinary figures.

The subspecies *asiatica* closely resembles the typical subspecies *marginata* already described, but differs from it in certain constant characters of which the most important are given below. [Not having seen the subspecies, it is proper that I should here acknowledge my indebtedness to the work of Moore (*loc. cit.*), which has been my only source of information.]

(1) The first pair of eyes are closely approximated, and so minute that they may easily escape observation. They generally lie immediately in front of the conspicuous and more widely

separated posterior pair of eyes, on the anterior ring of somite III, but sometimes are placed further forward and appear in somite II. In some cases they may be in contact with the posterior pair of eyes, and even apparently obliterated.

(2) The annulation in *asiatica* is considerably reduced. Somites I and II are uniannulate and III is also typically uniannulate, consisting of a broad ring, which, however, may show signs of subdivision dorsally, behind the eyes. Somite IV is biannulate and V is triannulate dorsally, its first and second rings fusing ventrally to form the free posterior edge of the anterior sucker. Somites VI–XXII are complete with three rings; XXIII is still triannulate, but with the last ring reduced; XXIV is usually biannulate, and XXV–XXVII are uniannulate and progressively smaller.

(3) Colour (in alcohol) reddish-brown, paler towards the extremities. The median series of spots often coalesce to form a more or less distinct longitudinal pale yellow band; the spots on the sensory rings, again, tend to spread laterally, forming broken or sometimes continuous pale yellow transverse stripes.

Dimensions and Habitat.—The size of the largest example is given as 16·3 mm. in length, with a maximum width of 6 mm. All the members of this subspecies are recorded from Kashmir, chiefly from slow-running streams.

Genus PARACLEPSIS, Harding, 1924.

Glossiphonidæ of medium size, with three pairs of eyes. First and second pairs on two consecutive rings, second and third pairs separated by two rings. Complete somite formed of three rings. Mouth-opening subterminal, leaving the anterior sucker imperforate. The crop (or stomach), has more than seven pairs of lateral diverticula.

22. Paraclepsis prædatrix, Harding, 1924. (Plate II, figs. 11 & 12; and fig. 36.)

The ovate-acuminate body is smooth below, but has a roughened or crustaceous dorsal surface due to numerous small papillæ closely set on every ring. A series of larger dorsal papillæ are present on the middle ring of each somite, and these consist of three pairs, occupying respectively a paramedian, an intermediate and a paramarginal position. The head-region is separated from the rest of the body by a slight constriction.

I am indebted to the late Dr. Annandale for the loan of a water-colour drawing of a living individual collected by him and also for the following description of its coloration :—" Semi-opaque pinkish-white, profusely ornamented with dull green pigment-cells on the dorsal surface. On the anterior and posterior thirds of the body these cells formed a broad and somewhat irregular longitudinal band, interrupted along the middle by a colourless line ; a transverse colourless line ran across the body just behind the

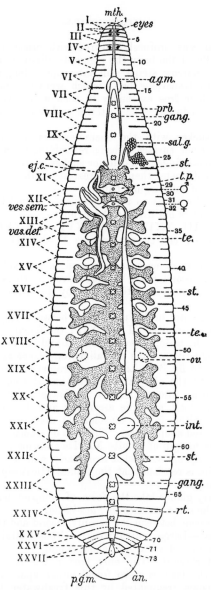

Fig. 36.—*Paraclepsis prædatrix*, Harding, 1924. Diagram showing external features, reproductive system, alimentary tract and ventral nerve-ganglia. Some of the paired organs are shown on one side only, in order to obtain clearness. Somites numbered in Roman, and rings in ordinary figures. *mth.* Mouth-opening. *prb.* Proboscis. *sal.g.* Salivary glands. *st.* Stomach. *int.* Intestine. *rt.* Rectum. *an.* Anus. *ej.c.* Ejaculatory canal. *t.p.* Its terminal portion. *ves.sem.* Vesicula seminalis or seminal reservoir. *vas.def.* Vas deferens. *te.* Testis. *a.g.m.* Anterior ganglionic mass. *p.g.m.* Posterior ganglionic mass. *gang.* A ganglion of the ventral chain. (After Harding.)

'head,' which was profusely covered with green pigment-cells. On the middle of the body the pigment-cells were arranged in a comparatively narrow stripe, still interrupted longitudinally, but giving rise to transverse bars, which were expanded and pro-liferated at their free extremity in such a way that a kind of network was produced. The posterior sucker bore faint, radiating green lines."

Anterior sucker with the characteristics of the genus and having its interior surface ribbed; resembling in this respect the tip of the human finger. Posterior sucker centrally attached and rather less in diameter than the greatest width of the body. It bears a series of submarginal papillæ.

Rings 73, of which two are preocular. Ring 71 is double at its margins, but is not divided throughout.

Somites I, III, XXIV, XXV, XXVI and XXVII biannulate; II uniannulate; the twenty somites IV–XXIII complete with three rings.

The three pairs of eyes are disposed in two subparallel rows. The first and second pairs lie respectively in rings 3 and 4; the third pair, separated from the others by the space of two annuli, lie in ring 7.

The male genital orifice is situated between rings 29 and 30, that is, between somites XI and XII; the female orifice lies two rings behind the male, between rings 31 and 32, in somite XII.

The reproductive organs and alimentary tract are represented schematically in fig. 36. Large sperm reservoirs connect the vasa deferentia with the ejaculatary canals and descend to about the fifteenth somite. Having reached its lowest point, each vesicula returns upon itself, and the ascending and descending portions are closely united for a considerable distance. The ovaries consist of paired simple sacs. The crop or stomach arises within the posterior margin of somite X. Its anterior portion expands bi-symmetrically but somewhat irregularly, and it is not until the twelfth somite is reached that the typical Glossosiphonid type of diverticula appear.

The salivary glands take the form of compact bunches closely resembling the same features in *Placobdella.*

The anus opens between the seventy-second and the last ring.

Dimensions.—Large individuals (in alcohol) attain a length of about 15·5 mm. and a width of about 4 mm.

Hosts and *Habitat.*—The only host noted is *Emyda granosa vittata,* and the leeches were found either upon their host or in ponds and pools frequented by this freshwater tortoise. The material, chiefly collected by the late Dr. Annandale, came from the following localities :—

(*a*) Tanjore, Trichinopoly District, S. India; (*b*) Bangalore, S. India, altitude circa 3000 ft.; (*c*) Kalka, at base of Simla Hills, altitude 2400 ft.; (*d*) Purulia, Manbhum District, Chota Nagpur Div., Bengal; (*e*) Selai Kusi, Magaldhai, Assam; (*f*) Igatpuri Lake, W. Ghats, Bombay Presidency.

23. Paraclepsis vulnifera, Harding, 1924. (Fig. 37.)

Description.—Body ovate-acuminate, with a somewhat rough-ened dorsal surface, due to the presence of numerous minute papillæ on every ring. (Accurate observations of papillæ and colour were prevented by the macerated state of the material.)

Head-region continuous with the body. Anterior sucker with the characters of the genus. Posterior sucker centrally attached, small, and less in diameter than half the greatest width of the body.

Rings 70. Rings 2 and 29 are double at their margins, but not entirely divided. Rings 7 and 8 unite below to form the first ventral annulus.

The first somite is uniannulate, and the second is represented by the anterior part of ring 2 containing the first pair of eyes. Somite III includes the posterior part of ring 2 together with ring 3. The twenty somites IV–XXIII are complete with three rings, and XXIV–XXVII are represented by the last 7 rings.

The three pairs of eyes are disposed in two subparallel rows. The small first pair are closely approximated in the anterior part of ring 2 (and may easily be overlooked); the second and larger pair lie in the posterior part of the same ring, and the third and largest pair are situated somewhat wider apart, in ring 5.

The male genital orifice opens between rings 27 and 28, that is, between somites XI and XII; the female orifice is separated by two rings from the male and lies between rings 29 and 30, the second and third rings of somite XII.

The reproductive and alimentary systems are shown in fig. 37. The large vesiculæ seminales seen in *P. prœdatrix* are absent.

The salivary glands and crop bear a close resemblance to the same features in *P. prœdatrix.* The crop arises in the anterior part of somite X. The anus lies between rings 69 and 70, being separated by one ring from the posterior sucker.

Size.—Length about 14 mm.; width about 8 mm. Living individuals probably attain a greater length.

Host and Habitat.—A note enclosed with the leeches states that they were taken from the branchial chambers of freshwater crabs (*Paratelphusa* sp.) at Mauganaltur, Tanjore District, Madras Province. They had been sent by Mr. Ballard, Government Entomologist, Madras Province, to Dr. Guy A. K. Marshall, C.M.G., Director of the Imperial Bureau of Entomology (Colonial Office), who was good enough to place the material at my disposal.

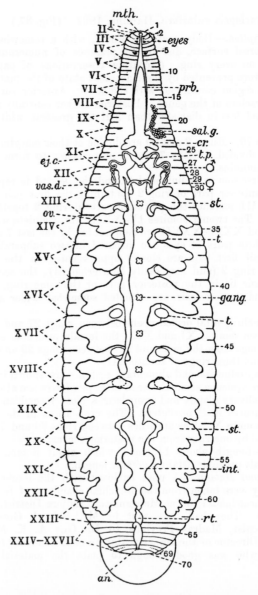

Fig. 37.—*Paraclepsis vulnifera*, Harding, 1924. Diagram showing external features, reproductive system and alimentary tract. Somites numbered in Roman, and rings in ordinary figures. *mth.* Mouth-opening. *prb.* Proboscis. *sal.g.* Salivary glands. *st.* Stomach. *int.* Intestine. *rt.* Rectum. *an.* Anus. *ej.c.* Ejaculatory canal. *t.p.* Terminal portion of ejaculatory canal. *vas.d.* Vas deferens. *Ov.* Ovary (one ovary only is shown, for the sake of clearness). *gang.* A ganglion of the ventral chain. (After Harding.)

SPECIES INQUIRENDÆ.

Family ICHTHYOBDELLIDÆ.

Genus BRANCHELLION, Savigny, 1822.

This genus is represented in the waters of both hemispheres, and although no species of *Branchellion* has so far (1926) been recorded from the Indian Region, it seems likely that such a record will eventually be made. The following diagnosis of this genus, therefore, will not be without interest here:—

Marine Ichthyobdellidæ parasitic upon fish. Body divided into two distinct regions : a short anterior "neck" and a long posterior "abdomen," with paired, lateral, foliaceous, non-digitate branchiæ and pulsating vesicles. Complete somite formed of three rings.

(?) Trachelobdella species.

Moore (1924, p. 306) refers doubtfully to the genus *Trachelobdella* a small leech in the collection of the Indian Museum which was not in a sufficiently good state of preservation for satisfactory determination. The species, which is stated to have been taken off the Pier, Ross Island, Andaman Islands, in 1915, is described as being nearly 10 mm. in length, with a very large posterior sucker, a wide, broadly ovate anterior sucker, and with traces of a pair of eyes and also of pulsating vesicles.

Family GLOSSIPHONIDÆ.

(?) Placobdella gracilis, R. Blanchard, 1897.

A single imperfectly preserved specimen of a leech in the Indian Museum collection, about 5 mm. long, found at Nandi, Mysore State, upon *Limnacea acuminata*, is referred with some hesitation to this species by Kaburaki (1921 (*b*), p. 702). The original examples of this little species described by Blanchard (1897, p. 334, fig. 2) came from Buitenzorg, Java, where they had been taken from the branchial chamber of a freshwater crab (*Paratelphusa* sp.).

(?) Placobdella parasitica, Say, 1824.

Oka (1922, p. 529) assigns "with much doubt" to this species a single, small and contracted specimen of a leech found upon *Taia shanensis*, Kobelt, in a canal on the western side of the Inlé Lake, S. Shan States, Burma. It was not deemed advisable to cut this single individual into sections, and after comparing it with the detailed description of *Glossiphonia* (*Placobdella*) *parasitica* given by Castle (1900, p. 51), Oka states that "it is certainly immature, and it is difficult to ascertain whether the slight but obvious discrepancies existing between this specimen and typical *P. parasitica* are due to difference in age or to specific distinctness."

BIBLIOGRAPHY.

APÁTHY, S. (1888 a.) Analyse de äusseren Körperform der Hirudineen. Mitth. Zool. Sta. Neapel, viii, p. 153, pls. viii & ix.

—— (1888 b.) Süsswasser-Hirudineen. Zool. Jahrb. iii, p. 725.

—— (1890.) Ertesitö az erdélyi Museum-egylet orvos-term. szakostályábol, xv, 1890, p. 110 (in Maygar), and p. 122 (in German).

BADHAM, C. (1916.) On an Ichthyobdellid parasitic on the Australian Sand Whiting (*Sillago ciliata*). Quart. Journ. Micr. Sci. lxii, pt. 1, n. s.

BAIRD, W. (1869.) Descriptions of some new Suctorial Annelides in . . . . the British Museum. Proc. Zool. Soc. London, 1869, p. 310.

BLAINVILLE, H. M. D. de. (1827.) Article "Sangsue" in Dict. des Sci. Naturelles, xlvii. 8°. Strasbourg and Paris, 1816–1830.

—— (1828.) Ibid. Article "Vers," lvii.

BLANCHARD, R. (1893.) Courtes notices sur les Hirudinées. X. Hirudinées de l'Europe boréale. Bull. Soc. Zool. France. xviii, 1893, p. 93. (*Placobdella.*)

—— (1894.) Hirudinées de l'Italie continental et insulaire. Boll. Mus. Zool. Torino, ix, 1894, No. 192.

—— (1896.) Hirudinées. Viaggio del dott. A. Borelli nella Argentina e nel Paraguay. *Ibid.* xi, 1926, No. 263.

—— (1897 a.) Hirudinées du Musée de Leyde, in 'Notes from the Leyden Museum,' xix, 1897, p. 73, pls. iv–vi.

—— (1897 b.) Hirudinées des Indes Néerlandaises. Zool. Ergeb. einer Reise in Niederlandisch Ost-Indien, Max Weber, Bd. iv, 1879, p. 332, fig. 1.

BOULENGER, G. A. (1904.) Cambridge Natural History, vii (Fishes), p. 613. 8°. Cambridge, 1904.

BRANDES, G. (1900.) Die Begattung von *Clepsine tessulata.* Zeitschr. f. Naturw., xxvii, pp. 126–128.

BRAUN, J. F. P. (1805.) Systematische Beschreibung einiger Egelarten. 4°. Berlin, 1805.

BRUMPT, E. (1900.) Reproduction des Hirudinées. Mem. Soc. Zool. France, xiii, pp. 286–430.

CASTLE, W. E. (1900.) Some North American Freshwater Rhynchobellidæ and their Parasites. Bull. Mus. Zool. Harvard, xxxvi, No 2, p. 147.

FILIPPI, F. de (1837.) Memoria sugli Anellidi della famiglia delle Sangui-sughe. 4°. Milano, 1837.

GODDARD, E. J. (1909.) Proc. Linn. Soc. N.S.W. xxxiv, 1909, p. 721, pl. lvi.

HARDING, W. A. (1909.) Note on Two New Leeches from Ceylon. Proc. Camb. Phil. Soc. vol. xv, p. 233.

—— (1910.) A Revision of the British Leeches. Parasitology, vol. iii, 1910, No. 2, p. 130, pls. xiii–xv.

—— (1920.) Fauna of the Chilka Lake: Hirudinea. Mem. Ind. Mus. vol v, No. 7, 1920, p. 510.

—— (1924.) Descriptions of some New Leeches from India, Burma and Ceylon. Ann. & Mag. Nat. Hist. Ser. 9, vol. xiv, 1924, p. 489, pls. ix–xv.

JOHNSON, J. R. (1816.) Treatise on the Medicinal Leech. 8°. London, 1816.

—— (1817.) Observations on *Glossopora.* Phil. Trans. Roy. Soc. 1817, p. 21.

JOHANSSON, L. (1896.) Ueber den Blutumlauf bei *Piscicola* und *Callobdella*. Festschrift Lilljeborg, pp. 317–330, pl. xvii. 4º. Upsala, 1896.

—— (1898 a.) Die Ichthyobdelliden im Zool. Reichsmuseum in Stockholm. Öfv. af K. Vet.-Akad. Förh. lv, pp. 665–687.

—— (1898 b.) Einige systematisch wichtige Thiele der inneren Organisation der Ichthyobelliden. Zool. Anz. xxi, No. 573, pp. 581–595.

KABURAKI, T. (1921 a.) On some Leeches from the Chilka Lake. Mem. Ind. Mus. v, 1921, No. 9, p. 661.

—— (1921 b.) Notes on some Leeches in the Collection of the Indian Museum. Rec. Ind. Mus. xxii, pt. v, 1921, p. 689.

KRØYER, —. (1850.) See Diesing, C. M. Systema Helminthum. (2 vols. 8º. Vindobonæ, 1850.) Vol. i, p. 438.

LAMARCK, J. B. de (1818.) Histoire naturelle des animaux sans vertébres.—V. 7 vols. 8º. Paris, 1818.

LINNÆUS, C. (1758.) Systema Naturæ, 10th ed.

—— (1761.) Fauna Suecica, 2nd ed.

—— (1767.) Systema Naturæ, 12th ed.

LIVANOW, N. (1902.) Die Hirudineen-Gattung *Hemiclepsis*. Vejd. Zool. Jahrb. (Abth. f. Syst.), xvii, pp. 339–362. 8º. Jena, 1902.

—— (1903.) Undersuchungen zur Morphologie der Hirudineen. I. Das Neuro- und Myosomit der Hirudineen. Zool. Jahrb. (Abth. f. Anat.), xix, i, pp. 29–90, Taf. 2–6.

MALM, A. W. (1860.) Svenska Iglar. Göt. Kongl. Vet. a. Vitt. Samh. Handlingar, viii, p. 153.

MOORE, J. PERCY (1898.) Leeches of the U.S. National Museum. Proc. U.S. Nat. Mus. vol. xxi, No. 1160, pp. 543–563, pl. xl. 8º. Washington, 1898.

—— (1900.) Description of *Microbdella biannulata*, with especial regard to the construction of the Leech Somite. Proc. Acad. Nat. Sci. Philadelphia, 1900, part i, pp. 50–73, pl. vi.

—— (1924.) Notes on some Asiatic Leeches . . . . principally from China, Kashmir and British India. Proc. Acad. Nat. Sci. Philadelphia, lxxvi, 1924, pp. 343–388, pls. xix–xxi.

MOQUIN-TANDON, A. (1826.) Monographie de la famille des Hirudinées. 4º. Montpellier, 1826.

—— (1846.) Ibid. Nouv. éd. with Atlas. 8º. Paris, 1846.

MÜLLER, O. F. (1774.) Vermium terrestrium et fluviatilium, i, Pars 2. 4º. Havniæ et Lipsiæ, 1773–1774.

OKA, A. (1894.) Beiträge zur Anatomie der Clepsine. Zeitschr. f. wiss. Zool. Bd. lviii, p. 79.

—— (1902.) Über das Blutgefassesystem der Hirudineen. Annot. Zool. Jap. iv, pt. ii, 1902, p. 49.

—— (1904.) Uber der Bau von *Ozobranchus*. Annot. Zool. Jap. v, pt. iii, 1904, p. 133.

—— (1910.) Synopsis der japanischen Hirudineen. Annot. Zool. Jap. vii, pt. iii, 1910, p. 165.

—— (1912.) Eine neue *Ozobranchus*-Art aus China (*O. jantseanus*). Annot. Zool. Jap. viii, 1912, pt. i.

—— (1922.) Hirudinea from the Inlé Lake, S. Shan States, Burma. Rec. Ind. Mus. xxiv, pt. iv, 1922, p. 521.

PHILIPPI, R. A. (1867.) Kurze Notiz über zwei chilenische Blutegel. Arch. f. Naturg., 1867 (I) pp. 76–78.

POIRIER et DE ROCHEBRUNE. (1884.) Sur un type nouveau de la classe des Hirudinées. C. R. Ac. Sci. xcviii, 1884, p. 1597.

QUATREFAGES, A. DE. (1852.) Études sur les types inférieurs de l'embranche-
ment des Annelés. Mémoire sur le *Branchellion* d'Orbigny. Ann.
Sci. Nat. xvii, Sér 3, 1852.

ROBERTSON, M. (1909.) Further Notes on a Trypanosome found in the
Alimentary Tract of *Pontobdella muricata.* Quart. Journ. Micr. Sci.
lvi, n. s. pt. 1, pp. 119–139.

—— (1910.) Studies on Ceylon Hæmatozoa. No. II. (*Hæmogregarina
nicoriæ*). Quart. Journ. Micr. Sci. lv, n. s. pt. 4, pp. 741–762,
pls. 32–41.

—— (1911.) Transmission of Flagellates living in the blood of certain
Freshwater Fishes. Phil. Trans. Roy. Soc. Series B, vol. ccii,
pp. 29–50, pls. 1–2.

SAVIGNY, J. C. (1822.) Système des Annélides .... in Description de
l'Egypte .... publié pas les ordres .... de Napoléon le Grand.
Fol. Paris. [The date of this volume is taken from C. D. Sher-
born's bibliography in Proc. Zool. Soc. London, 1897, p. 285.]

SCHMARDA, L. K. (1861.) Neue wirbellose Thiere, i, p. 2. Fol. Leipzig,
1861.

SELENSKY, W. (1906.) Zur Kentniss der Gefässystems der *Piscicola.* Zool.
Anz. xxxi, p. 33.

SUKATSCHOFF, B. W. (1912.) Beiträge zur Anatomie der Hirudineen. I.
Über den Bau von *Branchellion torpedinis,* Sav. Mitth. Zool. Sta.
Neapel, Bd. xx, 3, pp. 395–528, Taf. 18–24.

VEJDOVSKY, F. (1883.) Excrecni Soustawa Hirudinei. Sitzb. des Königl.
Böhm. Gesel. des Wissensch. Prag, pp. 35–51 and 1 pl.

VERRILL, A. E. (1872–1873.) Synopsis of N. American Freshwater Leeches.
Rept. U.S. Fish Commissioner for 1872–1873, pt. ii, p. 667.

WHITMAN, C. O. (1878.) Embryology of Clepsine. Quart. Journ. Micr. Sci.
xviii, n. s.

—— (1890.) Spermatophores as a means of hypodermic impregnation.
Journ. Morph., iv, pp. 361–406, pl. xiv.

—— (1892.) Metamerism of Clepsine. Festschr. zum siebenzigsten Gebur-
stage R. Leuckarts, pp. 385–395, pls. xxxix & xl. 4°. Leipzig, 1892.

Printed and bound by CPI Group (UK) Ltd, Croydon, CR0 4YY

21/10/2024

01777108-0001